NEW ENVIRONMENTAL POLICY INSTRUMENTS
IN THE EUROPEAN UNION

New Environmental Policy Instruments in the European Union

Politics, Economics, and the Implementation of the Packaging Waste Directive

IAN BAILEY

Department of Geography, University of Plymouth, UK

ASHGATE

Published by
Ashgate Publishing Limited
Gower House
Croft Road
Aldershot
Hants GU11 3HR
England

Ashgate Publishing Company
Suite 420
101 Cherry Street
Burlington, VT 05401-4405
USA

Ashgate website: http://www.ashgate.com

British Library Cataloguing in Publication Data
Bailey, Ian
  New environmental policy instruments in the European Union
  : politics, economics, and the implementation of the
  Packaging Waste Directive. - (Ashgate studies in
  environmental policy and practice)
  1. Packaging waste - European Union countries
  2. Environmental policy - European Union countries
  3. Environmental protection - European Union countries -
  Evaluation
  I. Title
  363.7'288

Library of Congress Cataloging-in-Publication Data
Bailey, Ian, 1965-
  New environmental policy instruments in the European Union : politics, economics, and
  the implementation of the packaging waste directive / Ian Bailey.
        p. cm. -- (Ashgate studies in environmental policy and practice)
  Includes bibliographical references and index.
  ISBN 0-7546-0888-3 (alk. paper)
    1. Environmental policy--European Union countries. 2. Environmental
  protection--European Union countries. 3. Environmental law--European Union countries.
  I. Title. II. Series.

GE190.E85B35 2003
363.7'056'094--dc21

2003043714

ISBN 0 7546 0888 3

Printed and bound in Great Britain by Antony Rowe Ltd.,
Chippenham, Wiltshire

Contents

List of Tables

List of Figures

Preface

Environmental policy in the European Union (EU) has undergone several transformations since its rather uncertain beginnings in the early 1970s. Although no mention was made of environmental matters in the Treaty of Rome establishing the European Economic Community in 1957, the EU has gradually consolidated the legal status of environmental policy to the point where it is now one of the core objectives of the EU and one of its most active areas of activity. Major changes have also taken place in the way environmental policies are implemented. European and national decision-makers initially relied almost exclusively on a command-and-control approach to environmental policy but in recent years there has been a rapid expansion in the deployment of more flexible instruments, such as environmental taxes, tradable permits, voluntary agreements and informational devices. This suite of 'new' environmental policy instruments emerged in part because it was felt that command-and-control techniques imposed unnecessary regulatory burdens and costs. It was hoped that more flexible policy tools would not only provide more economically efficient environmental protection but, in so doing, would also help to combat the growing implementation gap in EU environmental policy.

The aim of this book is to examine the contribution of these new policy instruments to the effective implementation of environmental policies in the EU. The underlying motivation for the volume is to progress academic debate on the practical efficacy of new environmental policy instruments to complement the extensive body of theoretical research in this area. It seeks to contribute to the debate on two levels. First, it evaluates the performance of new environmental policy instruments 'on the ground' in terms of their effects on business and market behaviour. Second, it assesses the politics behind the use of new environmental policy instruments in the EU.

The book is written to cater for a wide range of students, researchers and policy-makers but particularly targets readers seeking an introduction to practical and theoretical issues related to the implementation of EU environmental policy. Policy-makers and researchers will find the book useful for more detailed analysis of new environmental policy instruments. It is also recommended for environmental analysts within the business community and as a companion to other volumes in the Ashgate series on *Studies in Environmental Policy and Practice*.

The idea for this book grew out of research undertaken with the Department of Geography at the University of Plymouth and a workshop on new environmental policy instruments held by the European Consortium for Political Research in April 2001 and hosted by Institute of Political Studies in Grenoble, France. I am especially grateful to the workshop directors, Andrew Jordan, Rüdiger Wurzel and Anthony Zito, who, along with the other workshop participants, contributed immensely to my understanding of the politics of new environmental policy instruments. However, the project would never have progressed beyond an idea without the assistance of a great number of other people. First and foremost, I would like to thank Mark Blacksell and Mark Wise for their advice and guidance during the writing of this book and for their formidable appreciation of the labyrinths of EU politics. My sincere thanks also go to the Institute of Wastes Management for their financial support, to the UK Environment Agency for providing information used in the surveys, and to the hundreds of businesses and individuals that provided specialist information. Special mention here is due to the *Duales System Deutschland*, the *Arbeitsgrüppe Umweltstatistik* at the Technical University of Berlin, Perchards and Markus Haverland. Further thanks go to Richard O'Doherty and Alan Collins, with whom I collaborated in the writing of Chapter four.

I am also extremely grateful to David Pinder, Jon Shaw, Bettina van Hoven, Karl Martin Born, Erik Meyles, Francien van Soest, Roger Levermore and Jon Vernon for their comments and advice on earlier drafts. The technical and administrative staff at the Department of Geography in Plymouth also provided excellent support, including Tim Absalom, Brian Rogers, Maz Haji, Cheryl Hayward, Pauline Framingham and David Antwis. Finally, my profound thanks go to my wife, Rebecca, and daughter, Ruth, for being so supportive and for enduring academic loneliness with great fortitude.

Ian Bailey
Plymouth

Chapter 1

Introduction

1.1 Implementing European Union Environmental Policy: Themes and Issues

The credibility of the European Union (EU) environmental programme must ultimately be judged in terms of the effective implementation of its policies in the member states (Collins and Earnshaw, 1993). Yet weak implementation has been a characteristic feature of the environmental programme since its inception in 1973 and has, according to many accounts, seriously detracted from the advances made in the scale and scope of Community environmental legislation (Lowe and Ward, 1998a; EUROP News, 2000). As far back as the Fourth Environmental Action Programme in 1987, the Commission acknowledged that the implementation of EU environmental policies was unsatisfactory in many member states (Commission of the European Communities (CEC), 1987). In 1992, Ken Collins, then Chair of the European Parliament Committee on the Environment, concluded that: 'We have now reached the stage where if we do not tackle implementation and enforcement properly, there seems very little point in producing new environmental law' (House of Lords, 1992: para. 67). And more recently, Ludwig Krämer (1996: 7), Head of the Sustainable Development and Policy Instruments Unit in the Commission's Environment Directorate noted that: 'there are only a few areas of Community law in which the difference between the written law and the practice is as great as in the case of Community environmental legislation.'

Whilst numerous factors have been implicated as contributing to these problems (see Jordan, 1999 for a useful overview), the implementation of any environmental policy can be broadly viewed as a series of technical and political challenges. On the technical side, implementing mechanisms are needed that are capable of controlling environmental degradation without placing unreasonable restrictions on economic development and other aspects of social well being (Segerson, 1996). The political challenges facing EU environmental policy are equally complex because decision-making and implementation take place within a supranational political system that possesses many agendas but whose capacity for autonomous action is severely constrained by its founding treaties (Jordan, 1999). First, agreement amongst the member states is required for each new EU environmental policy as well as on the form of action to be taken. Second, protocols are needed for reconciling environmental policy with other areas of EU activity and to resolve any conflicts of priorities that might occur. Finally, monitoring and enforcement procedures are required to ensure that EU policies are effectively implemented in the member states (Howe, 1996; Haigh, 1998).

In terms of policy instruments, most EU governments have traditionally relied on legislation as the principal means for defining clear environmental standards for industry and the general public. In recent years, however, this approach has been increasingly criticised for its poor track record in arresting the general decline in the quality of Europe's environment (Pearce and Turner, 1990; CEC, 1992). Environmental economists in particular have proposed that 'traditional' legislation should be supplemented or replaced by so-called New Environmental Policy Instruments (NEPIs), such as eco-taxes, environmental charges, tradable permits (collectively termed 'market-based mechanisms' or MBMs), voluntary agreements, and informational devices (Jones, 1999)[1].

Theories on the use of MBMs and other NEPIs have developed at a remarkable rate in recent years with influential works such as Beckerman (1974) Baumol and Oates (1979; 1988), Pearce *et al.* (1989; 1993), Hahn (1989), and Tietenberg (1990). By refocusing traditional economics towards sustainability issues, environmental economists have proposed a range of fiscal mechanisms designed to promote effective and cost-efficient environmental policy (Repetto *et al.*, 1992; Stavins and Whitehead, 1992; Bahn, 1999). They suggest that environmental degradation is largely caused by the failure of market systems to place an adequate value on public goods, such as clean air and water. As a result, such resources tend to be over-used or their upkeep neglected by corporations in search of private profit. Economists suggest that this misallocation of resources can be corrected relatively easily through the creation of internalising charges in the form of eco-taxes and environmental charges (Folmer *et al.*, 2001). In simple terms, if pollution and resource depletion is made more expensive, polluters will be encouraged to adopt more sustainable patterns of production and consumption. Economists maintain that MBMs integrate environmental considerations more comprehensively into the economy than legislation, because the latter focuses principally on the creation of performance standards without giving sufficient consideration to the mechanisms and incentives needed for effective implementation. Furthermore, because MBMs are relatively non-prescriptive in terms of the types of action required, firms subjected to such instruments have greater flexibility in how they respond and are thus able to find cost-effective means of improving their environmental performance. Fiscal measures can also provide governments with additional tax revenue, which can be re-directed, or hypothecated, towards environmental or other re-distributive expenditures (Jacobs, 1991).

Tradable or marketable permits introduce a further innovation to market-based mechanisms, the exchange of pollution rights between firms within statutory emission ceilings set by government. They are based on the notion that any increase in pollution from one source must be offset by an equivalent or greater decrease elsewhere (Barde, 1997). Under tradable-permit schemes, if a company wishes to expand its polluting activities, it can only do so by purchasing permits from other firms that hold permits in excess of their requirements. This encourages businesses with low marginal abatement costs to reduce their pollution further, funded by trading with those that find abatement prohibitively expensive. The result is lower overall pollution and the use of market forces to reduce abatement costs.

Finally, voluntary agreements seek to encourage constructive partnerships between industry and government in environmental policy (Lévêque, 1995; Whiston and Glachant, 1996). They allow measures to be agreed between regulator and regulated in advance of implementation as well as encouraging the exchange of information and expertise required to define technically complex environmental standards. In return, by negotiating rather than imposing standards and time frames, voluntary agreements provide industries with firmer assurances that policy-makers are not overlooking their interests during the development of environmental policies.

For a variety of reasons, many EU governments were initially reluctant to do more than experiment tentatively with NEPIs (Huppes *et al.*, 1992; Howe, 1996; Helm, 1998). Also, many industry groups feared that NEPIs would add uncertainty and costs to their businesses and preferred the clear rules provided by command-and-control regulation. Progress was further inhibited by the legacy of existing regulation, which many politicians were reluctant to dismantle in order to experiment with 'new' instruments. In the United Kingdom (UK) in particular, powerful government departments like the Treasury eschewed such concepts as hypothecated environmental taxes because it was felt they increased the rigidity of the taxation system (Jordan *et al.*, forthcoming). Finally, despite the rapid ascent of environmental economics as an academic discipline, gaining a foothold in policy-making circles has involved challenging unsustainable institutions and ways of thinking that have been forged over many years (Pearce and Barbier, 2000). Creating strong advocacy coalitions for NEPIs has therefore proven problematic. This situation provided relatively few opportunities to undertake detailed empirical assessments of NEPIs that had been developed through 'laboratory-style' theoretical reasoning (see for example, Pearce *et al.*, 1989, 1993; Brisson, 1993; van den Bergh, 1996; Bohm, 1997; Ekins, 1997).

More recently, however, there has been a major increase in the use of NEPIs, particularly in developed countries. Pearce and Barbier (2000) note that the number of MBMs in the OECD (Organisation for Economic Co-operation and Development) countries increased from 100 to 169 between 1987 and 1992, with NEPIs being deployed to tackle such diverse problems as water and air pollution, waste management, fuel technology, energy consumption and air transport noise[2]. Some of the most significant recent developments in NEPIs have resulted from discussions on the implementation of the Kyoto Protocol on global climate change (Grubb, *et al.*, 1999). NEPIs under consideration as part of this programme include the Clean Development Mechanism, international emissions trading, voluntary agreements between the European Commission and automobile manufacturers, and the UK's Climate Change Levy (Keay-Bright, 2000).

The increased popularity of NEPIs can be attributed to a number of factors. First, there has been a growing acceptance of the limitations of regulation as a means of achieving environmental objectives and renewed interest in more innovative regulatory devices. Second, there has been a strong trend towards economic deregulation in many developed countries, partly stimulated by the recession of the 1990s, which has made the use of flexible and cost-effective environmental policy instruments more attractive to policy-makers. NEPIs were also seen as more consonant with the political ideologies of neo-liberalism and free-market economics that gained ascendancy in the 1990s and which have been reinforced by international pressure from the EU and the OECD (OECD, 1994). These forces have in turn prompted a greater cultural receptiveness to the ideas of environmental economics, particularly within the USA and the UK, where much of the pioneering research took place (Jordan *et al.*, forthcoming).

By corollary, the accelerated uptake of NEPIs has allowed greater assessment of their practical performance, particularly in the fields of waste management (Levenson, 1993; Fenton and Hanley, 1995; Defeuilley and Godard, 1997; Powell and Craighill, 1997; Sinclair and Fenton, 1997; Porter, 1998; Turner *et al.*, 1998) and air pollution (Baranzini *et al.*, 2000; Folmer *et al.*, 2001)[3]. However, these studies have mostly analysed national policies where NEPIs comprise only part of the overall regulatory package and have not always distinguished clearly the effects of individual NEPIs on polluter behaviour. As the principal advantage of these instruments is their use of market-based incentives to regulate industry's environmental performance, this lack of empirical investigation at corporate level is a significant omission from the literature. There have been some notable exceptions, such as the 'Porter hypothesis', which suggested that environmental regulations do not inevitably hinder the competitive advantage of firms and can enhance it (Porter, 1990; Porter and van der Linde, 1995) and Labatt's examinations of producer responses to environmental charges on packaging waste (Labatt, 1991, 1997a, 1997b). Nonetheless, the theoretical advances in this discipline have yet to be matched by comparable levels of empirical scrutiny.

Proponents of NEPIs have in fact themselves been at the forefront of acknowledging the theoretical bias in the discipline. Hahn (1989: 95), for example, confesses that the theoretical structure of environmental economics 'often emphasises elegance at the expense of realism,' whilst Jacobs (1991: 152) notes that the 'laboratory' models of NEPIs often: 'fail to represent the complexities of the real world, in which "institutional" factors crucially affect corporate and consumer decision-making.' More recent studies have also stressed the need for more applied work to inform theory and guide new policy initiatives (Baranzini *et al.*, 2000; Kaufman, 2001; Markyanda, 2001).

Political and institutional issues have also strongly influenced the implementation of the EU environmental programme. At first glance, the benefits of a common environmental policy are compelling. Many environmental problems are intrinsically trans-national in character and demand an international response. By pooling resources and expertise, therefore, concerted programmes can be developed and additional support provided to states that possess important environmental resources or poorly developed implementation capacity.

However, the reality of EU environmental policy often falls short of this ideal. Although the Union aspires to far-reaching economic and political integration, it remains a grouping of sovereign states that have formally ceded, in confederal fashion, certain policy-making powers to the EU institutions while retaining others at a national level (Wise and Gibb, 1993; W. Wallace, 1996). At the same time, political actors or appointees from the member states populate its key institutions and, thus, a neat separation of the pursuit of national interest from the supranational quest for common EU policies is fraught with problems. This creates tensions between, on the one hand, the political imperative to agree on important matters and, on the other, the desire to defend sovereignty and to have national agendas promoted on the EU arena. This is further accentuated by the presence of well-recognised environmental 'leader' and 'laggard' member states. The leader states – notably Germany, the Netherlands and the Scandinavian countries – are seen as a major force for the forward momentum of EU environmental policy. According to Sbragia (1996), this occurs where leader states introduce stringent national environmental standards and are keen to have them adopted across the Union in order to protect the competitiveness of their industries in European markets. The laggards are generally either sceptical about the scientific justification for precautionary environmental policies or concerned about the economic impact of ambitious initiatives. Because the issues and allegiances change with each policy proposal, environmental decision-making within the EU defies any simple classification but veers between 'lowest-common-denominator' bargaining, where vague and unambitious agreements result from the need to gain consensus, and a more 'entrepreneurial' style of decision-making seeking greater integration and higher environmental standards (Vogel, 1997; Zito, 2000).

Whilst these complexities pervade all aspects of EU policy, the tension between national and collective action is particularly apparent in the implementation of the environmental programme. For example, the European Commission is responsible for proposing new legislation to promote European integration and the specific objectives of the EU's Environmental Action Programmes (EAPs) but there are distinct limitations on its involvement in practical implementation. Under the Treaty, this remains the preserve of the member states, with the Commission serving an over-arching monitoring role. Thus, implementation is generally carried out at arm's length from the legislative process (Demmke, 1997; Haigh, 1998).

EU environmental law is usually enacted as a two-stage process. Assuming the member states can agree that European regulation is more appropriate than national action, the Council of Ministers and European Parliament must first approve the measures. Environmental legislation is then normally enacted in the form of directives, which are so-called because they 'direct' states to achieve particular objectives. Directives are therefore restricted to prescribing the obligation to act, the standards to be achieved and the time frame for compliance (Jordan, 1999). The member states then transpose the directive into national law and devise implementing mechanisms. Importantly, directives do not prescribe the methods national authorities must use to implement the policy. Whilst this arrangement is preferred by most member states because it makes EU law more sensitive to national circumstances and implementation capabilities, it has led to frequent disputes on the precise timing and extent of implementation required (Krämer, 1991). Moreover, because each state has its own perspective on how environmental policy should be prioritised and conducted – even within groups of states nominally classified as environmental leaders or laggards – the system has created considerable disparities in the way EU law is applied (Scott *et al.*, 1994; Knill and Lenschow, 1998).

This multi-level system of governance, with its intricate division of responsibilities and jurisdictions, has served to deeply politicise environmental policy in the EU. Although disputes over policy formulation most frequently grab the headlines, this politicisation also pervades the selection and deployment of instruments to implement EU policies. It should be said that these pressures are not unique to EU, as most political groupings are forced to confront similar issues when defining how policy goals should be pursued. However, they are certainly amplified by the supranational character of the EU and the distinctive policy processes that have emerged to manage the process of European integration. Often these political realities sit uncomfortably with the basic notion that, on virtually all accounts, the EU exists to solve problems that cannot be resolved at the level of the nation state (Weale, 1996). Similarly, EU activity is not restricted to environmental policy but spans a range of areas, including the Single Market and monetary union and, again, tensions are created because the boundaries of environmental policy are extremely difficult to define. In short, all forms of economic activity and development inevitably have some environmental consequences (Blacksell, 1994). The result is a highly complex and intriguing policy dynamic but one in which the interests of environmental protection are not always best served.

1.2   Aims and Objectives of the Book

Whilst there has been a great deal of academic interests in NEPIs and the environmental politics of the EU, the slow uptake of NEPIs by European governments means that a great deal of empirical scrutiny of their practical performance is still required. It must also be acknowledged that even the most sophisticated of theoretical models cannot to capture the full range of factors that influence market behaviour and, therefore, developing effective NEPIs is by necessity an iterative learning process. There has been a tendency for academic analysis to separate the normative economics of market behaviour somewhat artificially from the politics of environmental policy. However, within most political structures and in the EU in particular, these two dimensions are intrinsically linked. A key challenge for research, therefore, is to develop a more holistic assessment of the practical merits of MBMs and other NEPIs, which incorporates both technical and political factors, in order to understand better their contribution towards the effective implementation of EU environmental policies.

The central aim of this book is to assess how member states have deployed

NEPIs to promote the EU's environmental objectives in order to demonstrate the ways in which political (EU and national) and market factors have influenced their performance. General questions within this agenda include the following:

- How does the performance of NEPIs in practice compare with existing theory?
- What are the main environmental and economic advantages of NEPIs?
- What market factors influence the success or failure of NEPIs?
- To what extent do institutional influences (EU and national) affect the design and deployment of NEPIs and how do these impact on their performance?

In order to answer such questions and to highlight more general issues related to the implementation of environmental policy, the study analyses an EU policy where NEPIs have been widely employed. Pressman and Wildavsky (1984) employed this approach to great effect in their study of federal employment programmes in the United States. An alternative strategy would be to analyse the use of NEPIs across a range of different policies. However, because the operation and effects of NEPIs can vary considerably between policy areas, such 'breadth' studies may not adequately capture the complexities of what are often very detailed decision-making processes. Equally, the design of NEPIs can also be significantly affected by the nature of the environmental problem being addressed; again, this may hamper reliable comparisons. The strategy adopted in this volume, therefore, is to contrast how different member states have adopted and adapted NEPIs to address the same policy issue. As well as highlighting broader lessons for EU environmental policy, the findings will also, hopefully, stimulate further research into the practical applications of NEPIs. Whilst it is apparent that problems of comparison also exist with studies that compare the implementation of policies in different countries, where administrative traditions and practical circumstances may vary, this approach has the critical advantage of allowing the parameters of individual policy instruments to be examined under various institutional conditions.

In selecting an appropriate case study, temporal factors must also be taken into consideration. Because the effects of MBMs take time to filter through the economy, it would be unreasonable to judge a NEPIs too soon after its implementation. Assessment of the instruments deployed to implement the Kyoto Protocol on global climate change, for example, would be premature, since the first implementation deadline for this agreement is not until 2008-

12 and many of the implementing mechanisms to limit greenhouse gas emissions are still under discussion. It is therefore essential to select a policy where NEPIs are sufficiently well established that their performance can be reasonably assessed.

Based on these criteria, this book considers the case of the Packaging and Packaging Waste Directive (94/62/EC). The aim of the Directive is to harmonise national laws governing the reprocessing and disposal of packaging waste. It requires each member state to recover 50-65 per cent and recycle 25-45 per cent of its packaging waste (by weight) by June 2001 and also contains subsidiary objectives promoting the reduction, re-use and development of end markets for packaging waste (Official Journal of the European Communities (OJEC), 1994a: 12)[4]. Although the EU adopted this directive as a legislative instrument, it also sets out the conditions under which member states may use voluntary agreements and market-based instruments as part of their implementation strategies. Most national governments have consequently opted for a combination of legislative standards and NEPIs because they considered that legislation alone could not achieve the Directive's multiple objectives (Bailey, 1999a). The main NEPIs employed by the member states have been environmental charges, tradable permits, and voluntary agreements between government and affected industries. The Packaging Waste Directive has therefore emerged as a prominent example of the use of this array of 'new' policy mechanisms. Furthermore, as subsequent chapters will demonstrate, variations in the way member states have deployed NEPIs to manage packaging waste provide revealing insights into the relationship between NEPI design and the economic and environmental outcomes they produce.

To sum up, the book critically evaluates the negotiation, transposition and implementation of the Packaging Waste Directive in order to examine critical factors affecting the translation of NEPIs from theory into practice. As such, there are three key objectives:

- To establish, using the example of the Packaging Waste Directive, the manner in which NEPIs are used by the member states to implement EU environmental policy;
- To investigate the extent to which NEPIs have improved the implementation of EU environmental policy, with particular reference to their impact on industry;
- To assess the impact of EU decision-making and policy implementation structures on the efficacy of NEPIs in terms of environmental protection and the harmonisation of national policies.

In fulfilling these objectives, the study seeks to identify general lessons on the use of NEPIs in the EU. However, it is important to stress that because the study focuses on one particular piece of legislation, inevitably some of its conclusions will not apply to other areas of environmental policy. Similarly, member-state environmental policies are rarely static entities. Numerous adjustments to national policies have already occurred since this study began and such iterations are likely to continue with further developments in EU legislation. Indeed, at the time of writing, the Commission has recently proposed revisions to the Packaging Waste Directive to cover the period 2001-2006. The experience of the Packaging Waste Directive nonetheless provides a wide-ranging and dynamic example of how NEPIs are utilised to implement EU environmental policy.

Finally, it should be recognised that this volume only deals with one aspect of the implementation deficit in EU environmental policy. It is frequently argued that unsatisfactory implementation of the European programme exists at several levels: failure to introduce national legislation (transposition failure), failure to deploy implementing mechanisms (administrative failure), and under-performance of policy instruments against their stated objectives (practical failure) (Haverland, 2000a). In line with its key aims, this book concentrates primarily on practical implementation. Whilst transposition and administrative implementation are also considered during the course of this book, this is mainly to provide background information for assessing the practical performance of NEPIs. For more general overviews of the implementation deficit in EU environmental policy, see Collins and Earnshaw (1993), Krämer (1996) and Jordan (1999).

1.3   Structure of the Book

To address the issues highlighted in this introduction, the book is structured as follows. Chapter two examines the general aims of EU environmental policy and the 'toolkit' of implementing mechanisms available to policy-makers in order to provide a basic overview of EU environmental politics and NEPIs for the uninitiated reader. The discussion begins by outlining the evolution of the environmental programme from the United Nations Conference on the Human Environment in 1972 to its modern manifestation as a core EU policy committed, in rhetoric at least, to the concept of sustainable development (Ekins, 1993; O'Riordan and Voisey, 1998). It then reviews the politics of European environmental decision-making and implementation, where key features of EU environmental decision-making are highlighted. The chapter concludes by examining the various policy instruments used to implement environmental policy, paying particular attention to the relative merits of legislation and NEPIs.

The remainder of the book assesses the use of NEPIs to implement the Packaging Waste Directive. Chapter three traces the negotiation and transposition of the Directive and establishes how and why it became one of the most hotly contended pieces of legislation in the history of European environmental policy (Golub, 1996). In so doing, the chapter highlights how political tensions in EU environmental decision-making continue during the processes of formal and practical implementation in the member states. It then describes the implementation strategies adopted by two key protagonists in the policy debate, the UK and Germany. These countries were selected for detailed analysis because they have developed contrasting approaches towards the implementation of the Packaging Waste Directive (Haigh and Lanigan, 1995). Although both have employed ostensibly similar forms of NEPI, the British administration has attempted to nurture a market-led approach towards packaging recovery and recycling, which focuses strongly on mitigating the economic impact of meeting EU environmental standards. By contrast, Germany has developed a more command-and-control attitude towards the use of NEPIs in order to pursue ambitious environmental targets. NEPIs have therefore been deployed within starkly differing regimes of environmental governance, providing the opportunity to consider how prevailing institutional structures and political outlooks have shaped policy implementation in each country.

Chapters four and five assess the ways in which NEPIs have been employed to implement the Packaging Waste Directive. Chapter four reviews the development of recycling infrastructure and the incentives created by the use of voluntary agreements, economic instruments and tradable recycling certificates. It assesses the environmental and economic merits of each system and highlights the ways in which national authorities have modified NEPIs to achieve their desired policy objectives. Chapter five then analyses the effect of voluntary agreements and economic instruments on industries affected by the Directive. It begins by analysing the changes in waste management produced by the Directive, then assesses the extent to which these have been driven by economic instruments, voluntary agreements or legislation. These chapters draw upon correspondences and interviews with over 150 organisations involved in the management of packaging waste in Britain and Germany and surveys of 1800 businesses. On the basis of these results, conclusions are drawn on the nature of corporate responses to legislative and new environmental policy instruments.

Chapter six discusses the contribution of NEPIs towards the effective implementation of the European environmental programme and the convergence of member-state policies. First, it assesses the ways in which the EU's rules governing free trade have affected the design and deployment of NEPIs. Second, it examines the extent to which NEPIs are promoting the convergence or divergence of environmental policy implementation in the member states. Finally, Chapter seven examines the future prospects for NEPIs in EU environmental policy by assessing the implications of the Single European Currency, the Euro, and challenges posed by the enlargement of the EU to incorporate the former socialist states of Central and Eastern Europe.

Notes

[1] The limitations of legislation as a mechanism for implementing environmental policy are discussed by Schelling, 1983; Lévêque, 1995, 1996a; Skea, 1995; Whiston and Glachant, 1996. Following the groundbreaking work of Pigou; 1920, major reviews of economic instruments and other NEPIs include Baumol and Oates, 1988; Pearce *et al.*, 1989, 1993; Hahn, 1989; Tietenberg, 1990; Pearce and Turner, 1992, 1993; Helm, 1993, 1998; Turner, 1993; Turner and Pearce, 1993; Goddard, 1995; Turner *et al.*, 1998. Notable critiques of market-led environmental regulation include Daly and Cobb, 1990; Daly, 1992; Beder, 1996; More *et al.*, 1996.

[2] Data compiled by the Forum of the Future, Keele University, and the European Environment Agency.

[3] A number studies have examined the potential benefits of using economic instruments to regulate particular environmental problems, including Opschoor and Vos, 1988; Hahn, 1989; Jacobs, 1991; Goddard, 1995; Baranzini *et al.*, 2000. Several policy documents also assess the impact of NEPIs, for example, the Department of Trade and Industry (DTI)/Department of the Environment (DoE), 1991, 1992; Department of the Environment (DoE), 1993; Commission of the European Communities (CEC), 1994; OECD, 1994.

[4] Recovery denotes the collection of waste for recouping value, including incineration with energy recovery, recycling and composting. Recycling is defined as the reprocessing of waste materials excluding energy recovery. The packaging materials covered by the Directive are paper/cardboard, glass, steel/tinplate, aluminium, plastics, wood, and composites.

Chapter 2

EU Environmental Policy: Political Processes, NEPIs and Policy Implementation

PART 1 - ENVIRONMENTAL POLICY AND POLITICS IN THE EU

2.1　The Rise of European Environmental Policy

Although the EU has arguably one of the most progressive environmental policies in the world, the process of integrating environmental considerations into the wider fabric of European policy has been an extended one (Jordan, 2002). Environmental matters were ignored completely in the founding treaties of the European Economic Community (EEC) and were only formally incorporated into the EU's remit in the Single European Act (SEA) in 1986 (Blacksell, 1994). However, the first moves towards a European environmental policy began as early as 1972, following the United Nations (UN) Conference on the Human Environment in Stockholm. The conference was inspired by popular concerns about the state of the global environmental as well as influential studies such as the *Limits to Growth* report, which suggested that a combination of pollution, population growth and the exhaustion of natural resources was in danger of precipitating widespread environmental collapse (Ehrlich, 1971; Meadows *et al.*, 1972). Such dark prophesies and the UN conference's conclusions persuaded many European politicians that co-ordinated action was required to address common environmental problems and in 1973 the EU launched the first of a series of Environmental Action Programmes (EAPs).

　　Given the absence of a basis in the EU treaties, those involved in supporting the European environmental programme in the 1970s learned to operate very effectively in what was largely constitutional *terra incognita* (Jordan, 2000: 1309). This even proved to be a bonus in some cases, as the low political profile of environmental policy shielded it from excessive scrutiny and enabled an extremely active programme of regulation to develop (Lowe and Ward,

1998a). The early stages of the environmental programme have therefore been characterised as 'integration by stealth', with little serious contemplation being given to wider constitutional issues, such as the EU's competence to legislate on environmental matters or how far policy-making powers should be transferred from national governments to the EU institutions (Weale, 1999).

A stream of directives was generated in the First EAP (1973-6) covering air quality, water standards, waste disposal and noise pollution, many of which were inspired by existing Dutch and German legislation. In addition, a series of environmental principles were enunciated to guide the general direction of the environmental programme (see Table 2.1). Although these principles are not binding commitments on the member states in the same sense as the EU treaties or secondary legislation, they have formed the basis of several important rulings on the interpretation and application of EU environmental law. Examples of this in relation to the Packaging Waste Directive are discussed in Chapter three.

Whilst it is generally acknowledged that environmental policy prospered during the First EAP, the absence of formal legitimation in the treaties nevertheless meant that each intervention had to be carefully justified where it did not appear to have direct relevance to the EEC's core activities. The early stages of environmental policy therefore tended to be overwhelmingly technocratic, focusing on the development of tightly defined pollution standards in areas where the need for action could be readily defended. The early EAPs also concentrated mainly on the harmonisation of national environmental policies to remove trade barriers, though the environmental

Table 2.1   Principles affirmed in the EU environmental action programmes

a. Preventing pollution at source
b. Incorporating environmental considerations into all planning and decision- making
c. Adopting the polluter-says principle
d. Assessing the impact of EC policies on developing countries
e. Encouraging international co-operation
f. Promoting educational activities to increase environmental awareness
g. Ensuring action is taken at the most appropriate level (regional, national, EC)
h. Co-ordinating and harmonising the environmental programmes of individual member states
i. Improving the exchange of environmental information
j. Ensuring policies take a precautionary approach to environmental problems
k. The proximity principle; whenever possible, environmental damage should be rectified at source

Source: CEC (1984)

rationale behind these initiatives should not be overlooked (Barnes and Barnes, 1999). This focus remained largely unchanged throughout the Second EAP (1977-81), though greater emphasis was placed on international co-operation (Blacksell, 1994). Standards on water quality were strengthened and extended, as was the regulation of dangerous substances. The Second EAP also ventured into less trade-related areas, with directives covering habitats, endangered birds and other nature conservation issues.

The first real shift in emphasis came in the Third EAP (1982-6), where the integration of environmental considerations into other policy areas became a central concern (Baker, 1997). This reflected a realisation that environmental degradation could not be considered in isolation or controlled solely through technical pollution standards if environmental values were to be inculcated into all areas of EU activity. There was also renewed determination to bolster the implementation of environmental policy, as it became increasingly clear that the EU's vigour in producing new environmental legislation was not being matched in terms of enforcement on the ground (Barnes and Barnes, 1999).

The Fourth EAP (1987-92) marked a watershed in environmental policy in that it coincided with the SEA and the formal affirmation of the EU's legislative powers in respect of environmental policy. This new found confidence was manifested in the Fourth EAP by the adoption of a more holistic approach to environmental policy and by the Commission's decision to appoint an international task force to investigate the environmental impact of the Single Market, the result of which was the First Dobris Assessment. At the same time, the implications of transferring environmental decision-making powers to the European institutions were brought more sharply into focus after an extended lacuna during which most member states chose to turn a blind eye to the gradual extension of EU powers (CEC, 1984; Jordan, 2000).

The Fifth EAP came into operation in 1993 and charted a new, more reflective direction for environmental policy, founded upon the notion of sustainable development. The Dobris Assessment had drawn sobering conclusions on the first twenty years of EU environmental policy, highlighting a slow but relentless deterioration in Europe's environmental quality (CEC, 1992). The Commission went on to acknowledge in the Fifth EAP that existing policies had failed to deal adequately with the environmental problems caused by EU integration:

> The achievement of the programme and its objective of sustainable development constitutes one of the major political and economic challenges for the Community [and]...constitutes a major turning point (CEC, 1992: 145).

At its most basic level, the Fifth EAP sets out six issues as requiring special attention: sustainable management of natural resources, integrated pollution control and waste prevention, reduced consumption of non-renewable energy, improved mobility management, urban sustainability, and the improvement of public health and safety. In addition, five sectors were selected as the main foci of activities: industry, energy, transport, agriculture, and tourism (Blacksell, 1994). Finally, in an important shift in methodological emphasis, the Fifth EAP acknowledged that legislation alone – which had been the mainstay of EU environmental initiatives – would not be sufficient to achieve the long term ambition of sustainable development.

[I]n order to create market based incentives for environmentally friendly economic behaviour the use of economic and fiscal instruments would have to constitute an increasingly important part of the overall approach (CEC, 1992).

This landmark statement confirmed the EU's seal of approval for the accelerated use of more flexible instruments, such as environmental taxes and charges, tradable permits, and voluntary agreements in the member states. The Commission's change in outlook was also displayed in a slowing in the number of new legislative proposals and renewed emphasis on the effective implementation of existing measures (Jordan, 1999). Although the Commission maintained that legislation would continue to serve an important function, there was also a shift towards the consolidation of earlier legislation which had accreted in a rather piece-meal fashion during the earlier EAPs.

The Sixth EAP, entitled *Environment 2010: Our Future, Our Choice*, is more optimistic than its predecessor and seeks to commend the substantial progress made in environmental policy as well as reviewing ongoing problems and challenges (CEC, 2001a). Its action programme continues and extends many of the approaches of the Fifth EAP, including the emphases on sustainable development and flexible policy instruments, but identifies subtly changed priority areas: climate change, nature and biodiversity, environment and health, natural resources and waste. The inclusion of climate change and biodiversity in particular reflects the global environmental agenda created by the UN Rio Earth Summit in 1992 and its associated conventions. The Sixth EAP also advocates a more strategic approach to environmental policy based on wider constituency and the participation of all sections of society in the search for innovative solutions to environmental problems.

Summing up, environmental policy has emerged over the last thirty years from its uncertain beginnings to become a core element of EU activity. Over

500 legal measures have been introduced and the geographical coverage of the programme has extended massively as new member states have joined the EU. When the imminent accession of several Central and Eastern European (CEE) countries and trade agreements with non-member states are taken into account, it is only a slight exaggeration to say that the EU is now the driving force behind environmental policy across the majority of the continent. The environmental programme has also been transformed from its origins as a restricted body of technical standards designed primarily to eliminate trade restrictions into an expansive programme committed to the vision of sustainable development and the wholesale integration of environmental, social and economic policies. At the same time, the environmental programme is far from complete. Many important environmental indicators continue to deteriorate, constitutional frictions within the EU's supranational structure have become more evident and poor implementation has been an enduring, if not the defining, feature of EU environmental policy. In order to explore the reasons for these difficulties, it is necessary to examine the process of EU policy-making in greater detail. The following sections discuss the roles of the main EU institutions in environmental decision-making, the distinctive characteristics of these processes, and key issues influencing the development of environmental policy across the EU.

2.2   The Nature of European Environmental Decision-Making

For a number of important reasons, a considerable portion of the European politics literature is devoted to the complexities of environmental policy-making. First, because environmental policy was only formally recognised in 1986, well after the main frameworks of European integration were settled, it has been hemmed in by other Treaty requirements and forced to carve out policy niches in sometimes difficult circumstances (Lowe and Ward, 1998a). Second, the scope of environmental policy is never easy to define as all forms of economic activity and development inevitably have environmental consequences (Blacksell, 1994). The environmental programme has therefore entailed re-structuring other sectoral policies to take account of environmental pressures and, unsurprisingly, this has caused political frictions. Finally, despite the EU's initial neglect of environmental matters, the pervasive nature of many environmental problems militates against their resolution by nation states acting alone (Weale, 1996). Consequently, environmental policy provides a prime example of policy-making in an area where high levels of international co-operation are required but not always achieved (Keohane and Nye, 1989). At the same time, the very existence of a European environmental policy implies EU intervention in areas that have traditionally been the preserve of the member states; rules are therefore needed to demarcate between national and EU responsibilities.

Whilst many of the basic decision-making structures for environmental policy are outwardly similar to those in other areas of EU activity, unique pressures have influenced the nature of environmental policy-making. To preface the discussion, it is worthwhile first briefly identifying the main institutions involved in EU environmental decision-making and the roles played by groups outside the formal policy-making institutions, whose pecuniary or intellectual interests may encourage them to attempt to influence decisions through lobbying or more structured forms of engagement. Particularly within the environmental sphere, these typically include non-government organisations, business people, trade union functionaries and epistemic communities – networks of professionals with recognised expertise in a particular domain and an authoritative claim to policy-relevant knowledge (Haas, 1992). Together these comprise what are often termed policy networks, within which various representative bodies seek to maximise their power or influence over final decisions. By their nature, these policy networks are 'highly fluid, even evanescent,' since interests and issues tend to change even where policies only vary marginally from what has gone before (Walker, 2001: 276).

2.2.1 The EU Policy-making Institutions

The first EU institution with a major influence on environmental policy is the Commission, which Lister (1996: 10) describes as the 'regulatory engine' of the EU. Its principal function is to promote European economic and political integration through the development of proposals for new EU legislation. It is also largely responsible for monitoring member states' transposition and compliance with EU law (Cowgill, 1992). The Commission is divided into Directorates-General (DGs), each covering a specific policy area under the supervision of a Commissioner. However, although the Commission plays a pivotal role in formulating legislation affecting the citizens of Europe, it does not make final decisions on the acceptance of EU policies. Neither is it a democratically elected body; instead the DGs are mainly staffed by permanent civil servants whilst the Commissioners themselves are political nominees from each of the member states. Each DG is then divided into a number of services, with the Environment DG covering, amongst other things, Integration Policy and Environmental Instruments, Environmental Quality and Natural Resources, and Industry and the Environment. Since the Commission's responsibilities are divided between specialist departments, proposals emanating from one area inevitably impact on the work of other DGs and proposals must be co-ordinated within the Commission before they are formally submitted for discussion by the member states. This is particularly true of environmental policy because of the crosscutting nature of many environmental issues (Collins and Earnshaw, 1993).

The Council of the European Union (the Council of Ministers before the Amsterdam Treaty)[1] is the EU's main executive body on a wide range of Community issues and exercises legislative power in co-decision with the European Parliament. The Council is not a single entity of permanent representatives but is instead 'a revolving group consisting of the relevant ministers from each of the member states who meet periodically to decide upon Commission proposals which fall within their jurisdiction' (Lister, 1996: 15). Amongst its other duties, the Environment Council meets to debate, amend and adopt Commission proposals for Community environmental legislation. Golub (1996) clarifies the respective roles of the Commission and the Council by demarcating between the degree of *influence* and *power* held by each institution. Although the Commission has considerable influence over the environmental agenda because of its right to propose legislation, the member states exercise power through their participation in the decision-making executive, the Council. However, as the agenda is defined at least partially extraneously from the member states, this prevents the EU from being entirely sequestrated by national interests (also H. Wallace, 1996a). Tensions between the two institutions may arise where they hold differing perspectives on environmental issues *vis-à-vis* other policy areas. The situation is further complicated by the fact that the presidency of the Council rotates on a six-monthly basis. The agenda and short-term priorities of the environmental programme can therefore fluctuate dramatically within a relatively short time span, with the solution of problems being made significantly easier or more difficult by virtue of which member state holds the presidency at any particular time. The Committee of Permanent Representatives (COREPER) helps to maintain continuity within this system by preparing the work of the Council and carrying out tasks requested by the Council (Lister, 1996).

The third major institution is the European Parliament. Two principal powers were conferred on the Parliament in the Maastricht Treaty in 1992, the ability to propose and veto amendments to EU acts and the right, by two-thirds majority, to dismiss the Commission if it fails to fulfil its statutory duties (Cowgill, 1992). However, although the Parliament is the only directly elected EU institution, its influence over decision-making has historically been quite marginal. Weale (1999: 45) notes that:

Regarded from the point of view of parliamentary systems in Europe, the powers of the European Parliament appear few. It is not the formal source of legislation. It does not appoint or overthrow governments. Its party alignments are not well established. It is less attractive than national parliaments to those for whom politics is a career rather than a form of early retirement. It does not have the last say on legislative matters. In short, it still has to make the transition fully from a consultative body to a legislative body holding the executive to account.

Prior to the Single European Act the Parliament's role was almost entirely consultative. Under the consultation procedure, the Parliament was entitled to give legislation a single reading for the proposal of amendments but neither the Council nor Commission were obliged to accept Parliamentary suggestions. Amendments accepted by the Commission could be passed by a qualified Council majority but those which were rejected could only be adopted if they received unanimous Council support, though the Parliament retained the right to issue an official opinion on the final legislation (Wood and Yesilada, 1996). However, the Parliament's involvement in decision-making has been progressively enhanced by the adoption of the co-operation procedure in the SEA and the co-decision procedure in the Maastricht Treaty. Under the co-operation procedure, legislation rejected on its second Parliamentary reading can only become law if the Council unanimously over-rides the Parliament's veto (Figure 2.1). If the Council and Parliament fail to agree on a proposal which falls within the co-decision-procedure, a conciliation committee is formed to resolve their differences (Figure 2.2). As either party may reject the proposed solution, this effectively makes Parliament a co-equal legislative body in areas falling within this procedure. The Parliament's powers were further strengthened in the Amsterdam Treaty, where it was agreed that, with the exception of policies specifically exempted under Article 175, all environmental policies should be decided using the co-decision procedure contained in Article 251.

Although some commentators maintain that EU policies are still taken at some distance from direct democratic scrutiny, these extensions of the Parliament's powers have taken on additional significance in relation to environmental policy (Wood and Yesilada, 1996; Tsoukalis, 1997). The Parliament has assumed a particularly active role in this area, partly because of its strong 'Green' contingent and partly because environmental policy's broad scope has afforded the Parliament greater opportunity to utilise its powers under the co-operation and co-decision procedures. In particular, co-decision

Figure 2.2 The Co-desision Procedure

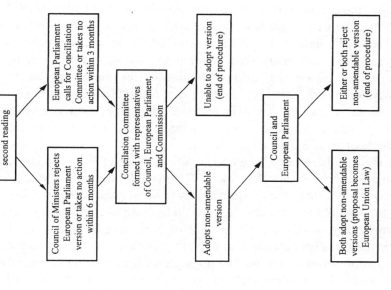

Source: Lodge (1993) cited in Wood and Yesilada (1996: 105)

Figure 2.1 The Co-operation Procedure

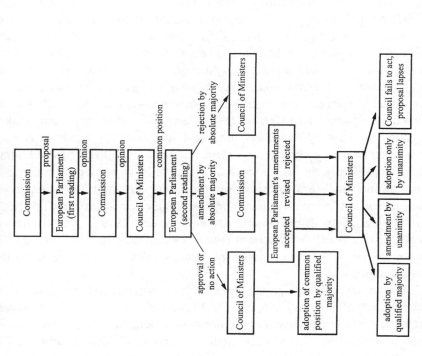

Source: McDonald and Dearden (1992) cited in Wood and Yesilada (1996: 104)

forces the Council to accept Parliamentary amendments it might otherwise wish to reject in order to avoid environmental initiatives failing entirely (Weale, 1999). The new procedures have therefore created an avenue whereby the European Parliament can extend its influence on environmental decisions beyond those customary for a national parliament. At the same time, both parties are aware that failure to adopt compromises proposed during conciliation committees is likely to result in the policy lapsing altogether. Co-decision has nevertheless increased the number of institutions and interests with significant influence over the policy process and, thus, has added to the complexities of EU decision-making.

The final EU institution with a significant bearing on environmental policy is the European Court of Justice (ECJ), which functions outside the formal decision-making process to provide legal safeguards for the founding treaties of the Community and the EU. Its judges are charged with ensuring that Community law is applied evenly in each member state and have the jurisdiction to hear disputes on the interpretation and implementation of EU legislation. Applications for ECJ proceedings against member states for failure to satisfy the requirements of EU law are normally instituted by the Commission as part of its enforcement remit and may include the granting of interim injunctions to prevent violations that are being contemplated or in progress (Lister, 1996). Whilst the Court is an important contributor to the enforcement of environmental law, it is also subject to certain constraints. For example, it must take into account precedents set in earlier rulings, whether environmental or otherwise and, therefore, developing a view of environmental policy that is entirely consistent with the broad aims of the EAPs can be problematic (Barnes and Barnes, 1999). Similarly, since the Court makes decisions on a case-by-case basis, individual judges with a jaundiced view of environmental policy may exert a considerable influence on rulings and future precedents. Finally, the judges can only accept reliable and accurate scientific information if they are to make informed decisions and, thus, imprecise concepts such as the precautionary and proximity principles can be difficult to incorporate into decisions (Barnes and Barnes, 1999). The Court has nonetheless fulfilled an important role in determining the direction of EU environmental policy and in ensuring that its rules are not flouted by the member states.

The creation of these separate jurisdictions was designed to avoid the over-concentration of authority in one EU institution and in recognition of the fact that the EU is a supranational entity not a fully-fledged state. However, this separation of powers inevitably creates a degree of tension between EU institutions and has marked implications for environmental policy. For example, whilst the Commission has the duty to pursue EU integration and the objectives of specific action programmes such as the EAPs, more disparate views are inevitably articulated within the Council because of the varied interests of the member states (Bulmer, 1983; Slater, 1982; Pfander, 1996). This friction between policy 'proposers' and 'deciders' is particularly pronounced in environmental policy because the Commission and the Parliament have traditionally been more sympathetic to environmental issues than the Council (Sbragia, 1996). Similarly, the co-decision procedure has reined in the powers of the Council to a degree but it has also placed new environmental initiatives in a more precarious position should either the Council or the Parliament be unwilling to compromise. It is also important to recognise that the diffusion of powers between EU institutions has inevitably created a bargaining mentality in EU decision-making, as individuals, institutions or member states seek to garner support for their viewpoints. Strategic bargaining is particularly commonplace in the Council, where member states may sanction proposals they do not fully support in exchange for concessions in areas where they have particular national interests (Lévêque, 1996b). The implications of this for EU environmental policy are examined shortly.

2.2.2 Non-Government Policy Networks

As noted earlier, the term policy networks covers a disparate ensemble of government and non-government groups or individuals that make or influence policy decisions. Typically, those parties from outside government circles include epistemic communities (natural and social scientists and professionals from other disciplines with a sufficiently strong claim to knowledge valued by policy-makers), environmental non-government organisations and industry groups (Walker, 2001). Haas (1989) argues that environmental policy-makers have increasingly courted the advice of epistemic communities to help in the elucidation of 'cause-and-effect' relationships and policy options to resolve particularly intransigent problems. Whilst Walker (2001) warns against unquestioning acceptance of the authority and objectivity of epistemic communities because of pronounced cultural differences in the way scientific knowledge is acquired and interpreted (see also Jasanoff, 1996), such groups have become important contributors in the formulation of environmental policies. One case in point is the influence of environmental economists, who have made a major contribution to the rise of NEPIs in many European states. At the same time, failure to heed such advice for political reasons has also been a characteristic trait of many EU environmental policies (Zito, 2000)[2].

Environmental non-government organisations (NGOs) have also increased their participation in European decision-making in recent years with the creation of the European Environmental Bureau in 1974. The main function of the Bureau is to liaise between individual NGOs and the EU institutions on environmental policy, with the intention of acting as a counter-weight to the industry lobby. By the end of the 1990s the EEB was a federation of 130 NGOs from 24 countries (Barnes and Barnes, 1999). However, maintaining the focus on the implementation of NEPIs, more detailed consideration is needed of industry's role in EU environmental policy, as industry has often been the main focus of these policy instruments. A number of authors have documented a marked rise in corporate lobbying at the EU (McLaughlin *et al.*, 1993; McLaughlin and Greenwood, 1995; McAleavey and Mitchell, 1994; Mazey and Richardson, 1993) and Tsoukalis (1997) even suggests that industry interests now dominate the formulation of environmental policies as the locus of decision-making has shifted from the national to the European level.

Corporate ventures into the EU arena generally take two forms: businesses can either oppose legislation that threatens their profitability or strategically support measures which offer potential competitive benefits (O'Brien and Penna, 1997). As such, both the *absolute* gains and losses of environmental legislation and their *relative* distribution between competing companies are important determinants of corporate response. This does not mean that businesses have the capability to veto environmental legislation, as this would require appreciable member-state or EU institutional support. In practice, they are more likely to adopt strategies which seek to minimise their absolute losses and maximise relative gains (van der Straaten, 1993). Lévêque (1996a) argues that, in pursuit of competitive gains, sequences of corporate lobbying frequently emerge. Companies threatened with the greatest absolute losses from an environmental initiative or those with more sophisticated monitoring and campaigning networks (generally speaking, large and multi-national companies) generally make the first attempt to influence policy-makers. This alerts other major companies, who realise they might become absolute losers should the first entrants secure a competitive advantage. This creates successive waves of corporate lobbying during which Commission proposals are progressively modified in order to maximise support. By contrast, the views of the Small-Medium Enterprise (SME) sector less frequently gain a full hearing since they lack the necessary networking capacity (Welford and Prescott, 1994). Lévêque's typology is framed around a divisive competition ethic but industries faced with common commercial threats (as often occurs with environmental legislation) may also form issue or sector coalitions to press their case (Jacquemin and Wright, 1994). National and cultural differences further complicate this mosaic. For instance, German industry has often demonstrated a general willingness to accept new social responsibilities provided they are introduced in a manner which does not disrupt competition (a relationship sometimes referred to as *Ordnungspolitik*), whereas British business tends to maintain a more individualistic and short-term view (Egan, 1997).

The intensity of industry lobbying is also affected by the price elasticity of products or services under policy scrutiny (Lévêque, 1996a). When faced with additional environmental costs for a price-elastic product (where demand may be significantly affected by an increase in prices), businesses are most likely campaign intensively against new regulation. Conversely, they may be more receptive to legislation where environmental costs can be readily recouped from customers without affecting demand for their products and services. In most cases industries will adopt the least *overall* cost response to environmental regulation. Determining the most favourable trade-off can be a complex process, however, since companies must assess whether it will be more expensive to absorb the costs of new regulation or mount an obstructive campaign. The publicity benefits of being seen to co-operate with policy-makers may also be an important factor (Welford and Prescott, 1994).

Despite the obvious conflicts between business interests and the EU environmental programme, the Commission has generally welcomed industry's input into environmental decision-making. One of the chief reasons for this is that the Environment Directorate has limited resources and expertise to define technically complex pollution legislation and therefore relies heavily on industry for information and advice. Such informationally-asymmetric relationships may render the Commission susceptible to regulatory capture, however, if companies use privileged data to add authority to their viewpoints or exploit scientific uncertainties to dispute the environmental risks associated with certain industrial processes (Böhmer-Christiansen, 1994).

Industry has also become instrumental in determining methods for implementing environmental policies. This has led to increased self-regulation, where industries voluntarily agree to control certain practices in order to stave off legislation, and co-regulation, where broad regulatory frameworks are established but industries are allowed considerable latitude in defining how environmental targets should be met. Self-regulation typically includes voluntary agreements while environmental taxes and tradable permits are sometimes seen as examples of co-regulation (Lévêque, 1995). The benefits and problems of these modes of regulation are discussed shortly.

In summary, industry's engagement with European environmental policy has been far-reaching but never straightforward. On the one hand, in attempting to influence policy-makers, industry representatives have themselves been influenced, helping to popularise the notion that constructive environmental management also makes good commercial sense (Gouldson and Murphy, 1996). Additionally, industry has used its technical expertise to good effect in developing innovative solutions to environmental problems. On the other hand, industry can be a powerful dissenting voice against policies it views as impractical or economically damaging and there can be few guarantees that industry's commercial interests will be consistently compatible with the aims of EU policy and sustainable development. Where such conflicts cannot readily be resolved through negotiation, they can re-surface with major implications during the practical implementation of environmental policies. Industry's engagement with EU environmental policy therefore defies neat classification though, in the final analysis, its onvolvement in policy formulation is an essential pre-requisite for the effective implementation of environmental policies in the member states.

2.3   Models of European Environmental Policy-making

2.3.1 Background

The European Union is a complex supranational entity and understanding its decision-making processes is a far from straightforward undertaking. Countless authors have analysed the dynamics of EU policy-making over the years and, doubtless, these debates will continue. The intention of this section is not to become involved in an extended discussion on the political science of EU integration but, rather, to elaborate the principal effects of EU decision-making on the environmental programme (for an accessible overview of more general issues related to European integration, see W. Wallace (1996)).

Whilst it is generally agreed that the EU does not yet possess the political attributes normally associated with a mature state, one of the central debates on European integration is the extent to which the EU remains an inter-governmental grouping of sovereign states or has been transformed into a more federal system of governance (W. Wallace, 1996; H. Wallace, 1996b; Tsoukalis, 1997). Inter-governmentalist explanations of European integration stress the importance of state independence, arguing that EU negotiations are typified by the defence of national interests and lowest-common-denominator bargaining. Despite the improved integration brought about by successive revisions to the EU treaty, inter-governmentalism has always formed a major part of European political relations (Slater, 1982). Moravcsik (1991: 216) remarks that:

From its inception, the EC has been based on interstate bargainings between its leading Member States ... each government views the EC through the lens of its own policy preferences; EC politics is the continuation of domestic policies by other means.

An alternative outlook on European politics is provided by confederalism, which acknowledges that the EU political system is not fully federal, as it does not possess a central executive with undisputed authority over its member states[3]. The member states have instead granted certain decision-making powers to the EU institutions whilst retaining others at the national level. However, confederalists argue that this cannot be equated to inter-governmentalism since these powers have been formally conferred in the treaties and member states cannot simply ignore EU decisions and processes. Although not all aspects of EU integration are necessarily compulsory – this was clearly demonstrated by the British, Danish and Swedish decisions not to join the single currency – clear rules of engagement exist to protect the Single Market and other areas of EU work.

Probably no single model can fully describe the evolving politics of the EU. Instead features of inter-governmentalism, confederalism and other models of European integration have each ephemerally characterised EU relations as political circumstances, issues and personnel have changed (Wise and Gibb, 1993). De Tocqueville (cited in Höreth, 1999: 249) even goes as far as to note that 'A new form of government has been found which is neither precisely national nor federal ... and the new word to express this new thing does not yet exist'. What is evident, however, is that tensions between national interests and collective action have a major bearing on EU environmental policies. Whilst similar institutional, ideological and interest pressures exist in all political groupings and policy areas, the trans-national dimension of environmental problems makes policy co-ordination a necessity rather than a luxury (Weale, 1996). However, transforming the ideal of collective action into reality can be an extremely problematic enterprise. As more parties become involved in decision-making – be they EU institutions, member-state governments or other players within a policy network, the more likely it is that the search for consensus will lead to *ad hoc* policy approaches, compromise solutions, and the exclusion of legitimate interests (Walker, 2001). Thus, the decision-making behaviour of the member states within the EU institutions has a crucial bearing on environmental policies and understanding these processes is a pre-requisite to understanding policy outcomes.

2.3.2   The Push-Pull Dynamic of EU Environmental Policy

Although the Commission is the main body entrusted with putting forward new environmental policies, relatively few proposals originate directly from the DGs (Collins and Earnshaw, 1993). Sbragia (1996) argues that, instead, much of the EU's environmental agenda is driven by a number of more environmentally progressive member states and their efforts to project their policies on to the rest of the EU. This, she contends, creates a 'push-pull' dynamic in environmental policy-making, where the internal politics of environmental leader states pushes new policies on to the European agenda and the EU's institutional framework pulls more sceptical member states towards levels of environmental protection they might not otherwise adopt (Sbragia, 1996). The process begins when one or more of the environmental 'leader' states introduces standards that are more stringent than those throughout the remainder of the EU. This would usually occur in response to domestic political pressures, particularly public support for environmental protection, and is most common in The Netherlands, Germany, Denmark and, since their accession in 1995, Sweden, Finland, and Austria (Krämer, 1991).

The pressure for policy 'Europeanisation' emerges where such laws to protect the environment also create a significant impediment to free trade in the internal market. In these circumstances the Commission comes under pressure either to challenge the national legislation or to propose harmonising legislation to remove the trade barrier (Bailey, 1999a). The leader state will usually be keen to see the latter course of action so that its national industries are not disadvantaged by the new environmental legislation. Pressure for standardised regulation may also come from businesses within the leader states. A notable example of this occurred when the German government introduced legislation to combat acid-rain problems caused by large combustion plants and asked the Commission to transpose its legislation word-for-word as a new Community directive (Zito, 2000). Although the Commission may still opt to begin proceedings if it feels the member state's law is unjustified, the chances are that this will result in lengthy and costly proceedings, particularly where national governments are able to defend their legislation by reference to the EU's own commitments under the EAPs (Lister, 1996).

Assuming the Commission accepts the case for EU legislation, the Council then debates the proposal. Depending on which area of the Treaty the proposal falls under, either unanimity or qualified majority support is required. At this stage 'laggard' member states and industries whose competitive position would be affected by the measures may oppose the proposal outright or seek to have its provisions diluted. Manifold justifications can be put forward by laggard state to resist the raising of environmental standards. For instance, the scientific evidence concerning the problem or its seriousness may be contested, the economic impact of higher environmental standards may be cited or constitutional objections may be raised to the EU's intervention in member-state affairs (Lévêque, 1996a, 1996b). At the same time, compromise is required in order to remove the potential distortion of the internal market. The EU therefore provides the justification and the forum for the Europeanisation of national environmental policies, with the Commission and the Parliament typically acting as policy entrepreneurs and the Council (increasingly in concert with the Parliament) as final decision maker (Weale and Williams, 1992; Weale, 1999). Regardless of whether lowest-common-denominator or more ambitious targets are adopted, the Council normally relaxes at least some of the Commission's original proposals. In order to remove any remaining trade distortion, the leader state may then be required to modify its original legislation to bring it into line with new European norms (Weale, 1996). A summary of the political actors and pressures that accompany the push-pull dynamic is provided in Figure 2.3. The cycle may then begin again, gradually strengthening the rigour of EU environmental policy and extending its influence across the member states. However, the push-pull dynamic is seen as a cumbersome way of conducting policy since it is very piece-meal and less oriented towards a problem-focused approach to sustainable development than to managing the democratic intricacies and multiple agendas of EU politics (Weale, 1996).

2.3.3 Consensual Bargaining and Policy-making by Concurrent Majorities

Prior to the SEA, all environmental policies required unanimous Council approval. Although environmental policy was highly active between 1973 and 1986, the requirement for unanimity was a noteworthy obstacle to the enactment of more ambitious policies, as individual states could automatically veto policies that displeased them (Lowe and Ward, 1998b)[4]. The principal innovation to streamline environmental decision-making in the SEA was the introduction of qualified majority voting (QMV) for measures concerning the Single Market. Under this system a proposal can be accepted if it receives 62 out of the 87 Council seats held by the member states (Barnes and Barnes, 1999). The division of voting rights across the EU institutions is shown in Table 2.2. The areas where QMV is applicable have been progressively expanded in the Maastricht and Amsterdam treaties such that they now cover the majority of environmental decision-making, with the following exceptions:

- Provisions primarily of a fiscal nature;
- Measures affecting land use and town and country planning, with the exception of waste and water resource management;
- Measures that significantly affect a member state's choice of alternative energy sources or the general structure of its energy supply (Barnes and Barnes, 1999).

Despite QMV, Collins and Earnshaw (1993: 225) note that strong political pressures still exist to achieve consensus in Council voting on major issues:

> Despite member states' articulation in Council of deeply entrenched preferences based on national circumstances and practices, negotiation in Council remains best characterised as a search for consensus...This search for unanimity...increases the possibility that EC environmental legislation will be vague, ambiguous and sometimes superficial.

In fact, still only 14 per cent of all Council decisions are made by QMV. Collins and Earnshaw remark that this can create a tendency for 'lowest-common-denominator' bargaining in EU decision-making that has clear analogies with the forms of inter-governmental decision-making normally associated with international environmental conferences. Weale (1996)

Figure 2.3 The push-pull dynamic of EU environmental policy-making

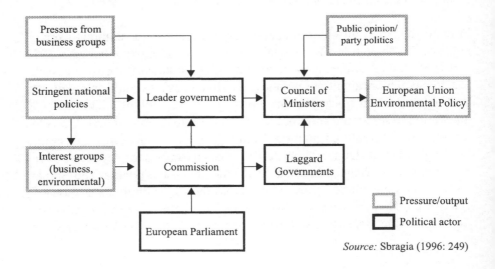

Source: Sbragia (1996: 249)

concurs that, within such a regime, policy *lourdeur* may result, with important measures not being adopted or only adopted in sub-optimal form. However, a number of important factors militate against this tendency. The first is the impetus provided by the EAPs and the EU's environmental principles, which set the broad agenda for more substantive legislation and against which success is measured at a strategic level. The presence of prominent pro-environmentalists in key EU institutions has also helped to propel standards beyond those normally characteristic of inter-governmentalism (Zito, 2000). A luminary in this respect was Carlo Ripa de Meana, the Environment Commissioner between 1989 and 1992, whose flamboyant style and willingness to champion radical ideas helped to create a dynamic period for EU environmental policy. Such influential figures have propelled EU environmental policy towards greater entrepreneurialism despite lowest-common-denominator bargaining remaining a prominent feature of European policy discourse.

Weale (1996) elaborates on this regime of governance and the policy networks it creates by describing EU environmental policy as decision-making

Table 2.2   Voting rights in the EU

Member State	Population (millions)	Parliament Seats	Council Votes	Commissioners
Belgium	10.2	25	5	1
Denmark	5.3	16	3	1
Germany	82.0	99	10	2
Greece	10.5	25	5	1
Spain	39.4	64	8	2
France	59.0	87	10	2
Ireland	3.7	15	3	1
Italy	57.6	87	10	2
Luxembourg	0.4	6	2	1
Netherlands	15.8	31	5	1
Austria	8.1	21	4	1
Portugal	10.0	25	5	1
Finland	5.2	16	3	1
Sweden	8.9	22	4	1
UK	59.3	87	10	2
TOTAL	375.4	626	87	20

Source: CEC (2000a: 63)

by *concurrent majorities*. He argues that environmental decision-making cannot be characterised either as a dominant coalition of member states consistently imposing its will on the minority or as a random 'merry-go-round' of individual countries seizing the agenda in areas where they have particular national interests. Instead, veto or obstructive power is sufficiently well distributed across the EU institutions that concurrent agreement is needed amongst a large number of participants before policies can be adopted. This requirement has become particularly noticeable since the advent of the co-decision procedure and the inclusion of the European Parliament as a significant player in the EU policy-making process. Dominant environmental policy-making is further precluded, Weale argues, by the fact that the issues and interests at stake tend to change with each environmental initiative.

Inevitably the need to gain concurrent majorities creates a tendency for environmental policy founded on negotiation and compromise, where final legislation is 'the aggregated and transformed standards of their original champions modified under the need to secure political accommodation from powerful veto players' (Weale, 1996: 607). The fundamental weakness in this structure, as with the push-pull dynamic, is that it encourages a bargaining mentality rather than 'objective' problem solving and creates an implicit conflict between the primary justification for EU environmental policy, the need for co-operation to combat trans-boundary problems, and the *praxis* of decision-making within the confederalist-intergovernmental framework. Even though environmental initiatives are not always weakened by the need to obtain concurrent majorities, the literature suggests that this is more often than not the case (Goodman, 1996; Bailey, 1999a; Jordan, 1999)[5].

2.4   Issues in EU Environmental Politics

2.4.1 Deciding on European Intervention: The Subsidiarity Principle

In terms of the functional stages of EU environmental policy, the first major issue is the definition of rules for sanctioning EU intervention in new areas of policy. Although many environmental problems are trans-boundary in character, it is by no means axiomatic that European policies will always be superior to national or local action. Many studies of sustainable development in fact suggest quite the opposite, arguing that excessively top-down policies are insensitive to local environmental and development needs (Morphet and Hams, 1994; Porritt, 1994; Agyeman and Evans, 1997). Equally, politicians in several member states have been alarmed at what they see as creeping federalism and the Commission's Europeanisation of environmental policy by stealth (Weale, 1999). The general principle used to demarcate between policy jurisdictions is that of subsidiarity, which is defined in the Maastricht Treaty thus:

> In areas which do not fall within its exclusive competence, the Community shall take action ... only if and in so far as the objectives of the proposed action cannot be sufficiently achieved by the Member States and can therefore, by reason of scale or effects ... be better achieved by the Community (quoted in Toth, 1994: 268).

At face value, subsidiarity would suggest a strong presumption in favour of decentralisation since it places the burden of proof on the European institutions to justify any moves to concentrate powers towards higher tiers of government (Jordan, 2000). Indeed, subsidiarity was conceived largely to protect local democracy during the EU's transition from purely economic collaboration towards political and economic union. However, as with all general principles, its interpretation can change with circumstances and the political lens through which it is viewed (Dovers, 1999). The imprecise terminology employed in defining subsidiarity criteria might even be seen through pro-federal eyes as inviting accelerated EU intervention. For instance, given the complexities of environmental policy, there are few objective means to decide whether a state can 'sufficiently' resolve a problem, whether solutions can be 'better' achieved by the EU, or what scale of intervention is most appropriate in particular cases (Green, 1994). From a legal perspective, therefore, subsidiarity may be virtually injusticiable since the Court of Justice would inevitably become embroiled in political rather than judicial decisions (Toth, 1994; House of Lords, 1996). Van Kersbergen and Verbeeck (1994: 220) encapsulate the difficulties surrounding subsidiarity, reporting the following caustic observation on its interpretation:

> The adoption of subsidiarity was cheered by both defendants of more authority at the Community level, like France and Germany, and opponents of such a development, as, for instance, the United Kingdom. Not surprisingly, subsidiarity rapidly became 'the Euroconcept all can admire by giving it the meaning they want.'

Subsidiarity has consequently become as much a symbol of the political problems associated with the EU's confederal-inter-governmental persona as it has a canon for allocating policy responsibilities. For some, subsidiarity, or a mechanism serving similar functions, is an essential pre-requisite of good

European governance (Blackhurst, 1994). Others maintain that it has merely smoothed over ideological differences 'by being so vague and insubstantial as to allow all parties to believe that it is furthering their cause, while in reality furthering none' (Green, 1994: 298). The emotiveness of these constitutional nuances may also divert political energies away from more material policy problems:

> A far greater service would certainly be rendered to the cause of environmental protection if instead of indulging in ideological disputes about interpretation of the subsidiarity principle the public debate concentrated much more on the serious causes of the shortfalls in implementation and enforcement and discussed the necessary reform of the environmental authorities and of environmental legislation (Demmke, 1997: 65).

Lister (1996) nonetheless notes that subsidiarity was formally elevated to a guiding principle of environmental policy under Article 130r(4) of the SEA, reflecting the widely-held beliefs that the EU had sometimes regulated too much and that some measures should be repealed or repatriated to the national level. The practical meaning of subsidiarity was clarified somewhat at the Edinburgh European Council in December 1992, where three operational rules were adopted (Jordan, 2000):

- The Commission should demonstrate that it has treaty-based competence to act before proposing new legislation;
- The Commission must demonstrate that the proposed action could not be achieved better through national or local action;
- The intervention should be kept to the minimum necessary.

The European Council further stipulated that the EU should only act when one of three 'tests' could be satisfied (Jordan, 2000: 1313):

- There are aspects of the issue which cannot be satisfactorily regulated by member states (the spillover test);
- Action by individual member states would distort competition in the internal market (the trade test);
- There are clear benefits by reasons of scale of effect for EU as opposed to national action (the added value test).

These tests were strengthened by a further protocol in the Amsterdam

Treaty. This required the Commission to justify its policy involvement using qualitative or, preferably, quantitative sustainability indicators, to base proposals on legislation which left member states with greater room for manoeuvre and, finally, to consult more widely and explain how its proposals comply with the requirements of subsidiarity (Jordan, 2000). The practical effects of this increased scrutiny have been a decline in the number of new environmental policies and a greater use of framework directives, which specify broad goals rather than detailed provisions (Barnes and Barnes, 1999). In the words of Jacques Santer, the Commission's new aim in environmental policy is 'to do less, better.' However, there is little evidence that subsidiarity has either shifted environmental decision-making significantly closer to European citizens or resolved the future constitutional role of the EU institutions in 'grey' areas of environmental policy (Jordan and Jeppesen, 2000). Despite nearly thirty years of experience, the allocation of policy responsibilities across of different tiers of government remains highly contested, which, in turn, has militated against the development of a consistent environmental strategy.

2.4.2 *Modes of Regulation*

EU actions can be communicated through a variety of media (decisions, recommendations and resolutions) but secondary legislation generally takes the form of either regulations or directives. Regulations are in effect direct EU law and are immediately applicable in all EU member states as soon as their terms provide without the need for national legislation (Lister, 1996). Directives are so called because they essentially instruct, or direct, member states to alter or supplement their national laws. They are more discretionary than regulations, as they are only binding in terms of the obligation to act, the overall objectives to be achieved, and the time frame for compliance. National authorities are then left to determine the detailed procedures for accomplishing the aims set out by the directive (Jordan, 1999). Directives are generally favoured over regulations in environmental policy because they allow national governments to employ legislative and implementation methods that are best suited to their country's circumstances. The member states have also proven reluctant to sanction the creation of an independent body of EU environmental law that may conflict with extant national legislation (Collins and Earnshaw, 1993). Finally, directives are designed to allow member states to introduce higher standards if they wish provided their measures do not obstruct free trade or other states' ability to comply with EU law. This enables environmental 'leader' states to respond to domestic pressures for more stringent legislation, thereby maintaining the momentum behind the push-pull dynamic (Krämer, 1991). The use of directives in environmental policy can therefore be viewed as an expression of the subsidiarity principle insofar as implementation decisions are left largely to the member states.

However, as a regulatory device, directives also have several weaknesses. First, they complicate the monitoring of national legislation against Single Market requirements, since more policy permutations are possible if implementation is managed by the member states (Collins and Earnshaw, 1993). Second, as directives are less prescriptive than regulations, their targets may be poorly defined, prompting disputes on the precise timing, nature and extent of implementation (Demmke, 1997). This problem is compounded by the fact that infringement proceedings are lengthy and, even if the Court of Justice finds against the state, the Commission has limited powers to enforce rulings because it cannot force national or sub-national authorities to take corrective action. The Commission has therefore tended to use court proceedings only when it has exhausted other diplomatic channels (Jordan, 1999). Finally, because directives do not prescribe implementing methodologies, member states have frequently used this flexibility to re-interpret EU law in ways that fundamentally alter the environmental outcomes achieved by ostensibly common policies (Bailey, 2000). The implications of this for the implementation of the Packaging Waste Directive are examined further in Chapters three and four. As with subsidiarity, therefore, modes of regulation have evolved as much to deal with constitutional issues as to manage environmental problems, while the complexities of multi-level governance have added to the range of potential conflicts created by the crosscutting nature of environmental issues.

2.4.3 The Under-Implementation of European Environmental Policies

The implementation of EU environmental law has been described as the poor relation of policy analysis despite it being arguably the weakest element of the environmental programme (Demmke, 1997). In many respects, this may itself be symptomatic of a rather stylised view of the EU, based on the assumption that once the important decisions have been taken by the supra-national institutions, national authorities will faithfully transpose and implement these policies. However, it has become increasingly apparent that during practical implementation the tension, or dualism, between the intergovernmental and confederal aspects of the EU is most starkly revealed, as it is here that the idea of collection action has to be physically reconciled with the system of state-dominated implementation (Jordan, 1999: 77).

Demmke (1997) argues that implementation problems often begin during the drafting of EU legislation, as many texts are of poor legal quality and establish vague, or even contradictory, objectives. Coupled with this, European directives tend to cut across national laws and jurisdictions, which are themselves highly complex and heterogeneous (Collins and Earnshaw, 1993). This problem is especially acute where responsibilities for overseeing environmental policies have been partially or fully devolved to sub-national governments, as is the case in Germany, Belgium, Austria, and, increasingly, Spain and the UK. It is also argued that issues pertinent to implementation are not given fullest consideration in the drafting of legislation (Jordan, 1999). One explanation for this is the fact that the body responsible for proposing legislation, the Commission, is not substantively involved in practical implementation. Notwithstanding this, whilst there are variations in the way member states transpose EU environmental directives into national law, non-compliance in terms of failing to enact relevant laws is not necessarily the most pressing problem. The record for most member states in this regard might even be considered quite reasonable (Table 2.3).

Of greater concern has been member states' shortfall in practical implementation and failure to meet required environmental standards (Table

Table 2.3   Progress in transposing environmental directives

Member State	Directives applicable, 31 December 1997	Directives for which measures notified	% transposition
Belgium	139	121	87
Denmark	139	139	100
Germany	141	133	94
Greece	144	140	97
Spain	143	142	99
France	139	133	96
Ireland	139	136	98
Italy	139	135	97
Luxembourg	139	136	98
Netherlands	139	137	99
Austria	135	131	97
Portugal	143	138	97
Finland	137	132	96
Sweden	137	133	97
UK	139	133	96

Source: CEC (1998c: 95-96) in Barnes and Barnes (1999:100)

2.4). Some implementation failures can be attributed to genuine difficulties in matching EU and national laws. The implementation of the directive on the Quality of Water intended for Human Consumption (80/778/EEC), for example, required changes in seven areas of national law in the UK (Barnes and Barnes, 1999). In other instances, member states have been reluctant to adopt the necessary implementing measures. For example, Germany has been accused of dragging its heels on the implementation of the Access to Environmental Information Directive (90/313/EEC), which obliges member states to make publicly available environmental information held by public authorities. Germany has resisted this directive in part because there is no tradition of general right of access to environmental information in German law and, thus, a clash of administrative cultures has occurred (Kimber, 2000).

A final cause of unsatisfactory implementation is the difficulties in maintaining effective enforcement procedures, both at national and EU level. The member states have highly varied enforcement structures in terms of resources, remit, and level of independence from government authorities; some national agencies are therefore better equipped to monitor compliance in their territories than others (Demmke, 1997). Equally, the Commission has extremely limited powers to conduct direct assessments of member states' compliance with environmental law. Although the European Environment

Table 2.4   Practical infringements of EU environmental law in 2001

Belgium	12
Denmark	4
Germany	14
Greece	13
Spain	23
France	21
Ireland	14
Italy	14
Luxembourg	8
Netherlands	9
Austria	6
Portugal	19
Finland	5
Sweden	3
UK	9
TOTAL	174

Source: CEC (1999a: various)

Agency (EEA) was established in 1990 to assist the implementation process, it differs markedly from national environment agencies in that its function is 'largely informational rather than regulatory or implementational' (Lister, 1996: 15). Proposals to transform the EEA into an international inspectorate have been roundly rejected.

Should a member state fail to transpose or properly implement a directive, first responsibility for enforcement falls upon the Commission. Initially this takes the form of bilateral exchanges with the member-state government to resolve outstanding problems without recourse to formal proceedings. Should these fail, the Commission informs the state in a '226 letter' that it believes a failure to fulfil Treaty obligations has taken place[6]. The letter also specifies a deadline by which the state's observations are required (Collins and Earnshaw, 1993). Though relatively few proceedings progress beyond this point, the Commission may issue a 'reasoned opinion' if it is not satisfied with the state's reply. This sets out the reasons why the state's justifications are not accepted and a timeframe for compliance. Where a Member State persists with a transgression, the Commission may apply to begin proceedings with the ECJ. Whilst receiving a 226 letter is normally enough to shame a member state into action, the number of infringements of EU environmental policy continues to rise year on year (Jordan, 1999; Environmental Data Services (ENDS), 2000a). Furthermore, if a member state ignores a Court ruling, the Commission is relatively powerless because it has no direct authority to force the state to take action.

One of the more significant developments in enforcement at the EU level

Table 2.5   Examples of requests for penalty payments

Member State	Subject	Penalty payment (ECU/day)	Date of decision	Settled
Italy	Radiological protection	159,300	29 January 1997	Yes
Italy	Waste management plan	123,900	29 January 1997	Yes
Germany	Surface water	158,400	29 January 1997	No
Germany	Wild birds	26,400	29 January 1997	Yes
Germany	Groundwater	264,000	29 January 1997	Yes
Belgium	Wild birds	7,750	10 December 1997	Yes
Greece	Waste water management	24,600	26 June 1997	No

Source: CEC (1998a: annex III)

has been the introduction of a system for fining member states that repeatedly defy ECJ rulings (Article 171). In 1997 the Commission recommended seven cases where it believed member states should receive daily fines for persistent infringements of EU environmental law (Table 2.5). The Court imposed the first actual fine on Greece in July 2000 for a second breach of a ruling and the Commission plans to apply for further fines against Germany and Britain in the near future (ENDS, 2000b)[7]. It is probably too early to make a detailed assessment of whether fines will substantially improve the practical implementation of EU environmental law. However, their introduction represents a landmark in EU environmental politics because it signifies that the member states have agreed that non-implementation of environmental policy is an issue that warrants concerted action. It should also be noted that less confrontational initiatives have been introduced to address the implementation issue. These include the Access to Environmental Information Directive, which is designed to increase public participation in monitoring, LIFE (*l'instrument financier pour l'environment*), which provides financial assistance in developing implementation capacity, and several consultative fora introduced to enhance dialogue on implementation issues and the transfer of good practice. Despite these initiatives, practical implementation remains the Achilles heel of the EU environmental programme and one which the EU's multi-tiered system of governance has struggled to combat effectively.

2.5   Concluding Comments

In drawing up a balance sheet of the successes and failures of EC environmental policy, the outstanding feature is that it exists at all (Blacksell, 1994: 341).

Considering the inherent conflicts between environmental policy and other areas of EU activity and the fact that environmental issues were not even mentioned in the Treaty of Rome, it is remarkable that this area of policy has attained its current level of prominence. Not only has the number of environmental measures increased immensely between the 1970s and the new millennium, the scope of EU environmental policy has also been transformed from a rather motley assortment of trade-related technical standards into something approaching a comprehensive framework for achieving sustainable development. Against this, the EU's efforts to reverse environmental decline in Europe have enjoyed only partial success. Disputes over when to permit EU intervention in national affairs have never been entirely resolved despite subsidiarity and, even where policies have been agreed, the ideal of synergetic collective action continues to sit uneasily alongside member states' desire to maintain control over important aspects of decision-making. Although EU environmental policy has now largely moved beyond its early focus on intergovernmental and lowest-common-denominator bargaining, remnants of these decision-making modes have continued to hinder progress. Whilst it is inevitable that clashes between competing ideologies and national perspectives will occur in any international forum on environmental issues, leading to compromise and 'sub-optimal' policies, this has been compounded in the EU by the deficits in policy implementation that have emerged in the member states. Whether or not such issues might be resolved or further exacerbated by the move towards NEPIs remains an open question and is discussed in later chapters.

PART 2 - ENVIRONMENTAL POLICY INSTRUMENTS

2.6 Introduction

At first glance, the cut and thrust of European environmental politics seems far removed from the scientific study of policy mechanisms for mitigating environmental problems. Much of the pioneering work on new policy instruments has been conducted by economists, whose natural predisposition is towards the advancement of knowledge in what might be termed cognitive laboratories. In recent years, however, policy-makers' eagerness to experiment with NEPIs has increased economists' engagement with implementation issues, allowing theory and practice to cohere more effectively (Pearce and Barbier, 2000). Correspondingly, the politicisation of NEPIs has also increased, as governments have sought to create portfolios of policy instruments that meet political and economic ends as well as environmental objectives (Jordan *et al.*, forthcoming). During this time governments have undergone a steep learning curve in which the deployment of NEPIs has been informed by a combination of scientific evidence, persuasive policy networks (including epistemic communities), political and economic ideology, and practical exigencies. Within such a climate of methodological advancement and politicisation, the resulting policies and instruments can sometimes seem rather muddled and confused (Helm, 1998).

Chapters three to seven examine the way in which political and economic factors have influenced the design and application of NEPIs used to implement the Packaging Waste Directive and other EU policies. In order to provide a context for this discussion, the remainder of this chapter outlines basic theories on the design and deployment of NEPI. As with the discussion on EU environmental politics, the intention is to avoid an extended and distracting theoretical discussion. Instead the focus is on providing an introduction to the defining features of each key mode of environmental regulation; command-and-control regulation (legislation), environmental taxes and charges, tradable permits, and voluntary agreements. In each sub-section, the basic purpose of the policy instrument is identified then a critical commentary is provided on its strengths and weaknesses.

2.7   Command-and-Control Regulation

Most European governments have traditionally relied on legislation as the mainstay of environmental policy. Such regulation is typically expressed in the form of command-and-control emissions standards, licences, prohibitions or requirements to employ particular technologies[8] (Hahn, 1993; Börkey and Lévêque, 2000). The principal attraction of legislation is that it offers reasonable certainty as to the end result, since clear environmental standards and responsibilities can be created to prevent or limit environmental damage (Segerson, 1996). However, the effectiveness of legislation is dependent on, amongst other things, companies complying with legislation and, thus, the legal and administrative structures which uphold the standard (Department of the Environment (DoE), 1993). Where governments hold a strong degree of coercive power over industries they wish to regulate, compliance with environmental legislation will be high. Lévêque (1995) argues that the effectiveness of legislation also depends on the extent of private incentives for business to change their behaviour, the level of informational asymmetry between regulators and regulated, and the degree to which companies engage in opportunistic behaviour, such as strategic non-compliance. The latter may predominate if the penalties for non-compliance are less than the advantages to be gained from infringing government standards.

There have nonetheless been concerns that command-and-control techniques have not coped with longstanding environmental problems and have not integrated environmental considerations adequately into mainstream economic and planning decisions (Ekins, 1999). In the Fifth EAP, the Commission noted bluntly the failure of case-by-case legislative instruments to arrest the decline in the quality of Europe's environment, arguing that they had not ingrained environmental protection firmly into the cost considerations of the market and, therefore, the minds of business executives and consumers (CEC, 1992). There is also intense debate about the economic ramifications of standards-based legislation. Although legislation does not impose direct charges on polluting activities, costs are obviously incurred by industry in setting up procedures to meet standards and by public authorities in monitoring compliance (Beder, 1996). Indeed, some see legislation as unnecessarily expensive, arguing that uniform standards contrived by government officials who are remote from, and ill informed about, market conditions are insensitive to the needs and capabilities of individual companies and sectors (Baumol and Oates, 1988). In the absence of conclusive scientific evidence on the extent to which environmental degradation can be attributed to particular pollutants or actors, policy-makers may also impose over-burdensome and ill-targeted standards (Helm, 1998). The precautionary principle may provide general guidelines but is more of a moral injunction than a precise policy instrument (Dovers *et al.*, 1996). Whilst debate on the efficacy of command-and-control regulation is ongoing, the perceived weaknesses of legislative instruments has led to heightened academic and policy interest in alternative environmental policy instruments. Many economists (for example, Baumol and Oates, 1988; Pearce *et al.*, 1989; Jacobs, 1991; Ekins, 1999; Folmer *et al.*, 2001) have proposed that in many areas (but not all) NEPIs can be deployed to achieve equivalent or higher environmental standards at lower cost than traditional legislation.

2.8   Environmental Taxes and Charges

The basic rationale behind environmental taxes and charges is the use of fiscal instruments to correct environmental externalities, the negative impacts on environmental quality caused by the failure of markets and production processes to consider the side-effects of their activities (Pearce *et al.*, 1989; van den Bergh, 1996). Externalities arise where private individuals or firms only consider the private use value of environmental resources and neglect their option and existence values as well as the costs to society caused by

Figure 2.4 Valuation of environmental resources

Total Economic Value = Actual Use Value + Option Value + Existence Value + Pollution/Pollution Avoidance Value

where:

Option Value = Value in Use (by the individual) + Value in use by future generations + Value in use by others (vicarious value to the individual)

Source: Pearce *et al.* (1989: 7)

different forms of pollution (see Figure 2.4). The incentive for so doing is the prospect of receiving the profit derived from polluting the environment – which consumers may also share in the form of cheaper goods and services – whilst the costs of environmental neglect are borne by society as a whole. By levying taxes or charges on activities that create adverse externalities, the cost of excessive environmental degradation can be re-internalised into the market, encouraging more prudent resource-use or pollution patterns. The primary function of these instruments, therefore, is to create a financial incentive for reducing pollution, based on the polluter pays principle (PPP) or its latter-day variant, the user pays principle (UPP) (Fenton and Hanley, 1995). Ekins (1999) points out, however, that environmental taxes and charges can serve a multitude of (not necessarily exclusive) purposes and distinguishes between their incentive and revenue-raising properties.

Cost-covering Charges
This form of economic instrument raises levies from polluters principally to cover the costs of providing specific environmental services, such as the recycling of waste or the administrative costs of regulation. Alternatively, cost-covering charges can be earmarked, or hypothecated, for related environmental expenditure not classified as a specific service to the charge payer. The level of the charge is determined by the cost of the service and the instrument might not be designed with any particular ambition to change polluter behaviour (Ekins, 1999).

Incentive Taxes
Environmental taxes may also be raised with the intention of changing polluter behaviour without having a specific revenue-raising intent. The level of such taxes is determined by calculating the marginal cost of environmental damage and the marginal economic value derived from that damage (Folmer *et al.*, 1995). The tax is considered to be optimal when it is set at a rate where the marginal cost and benefit are equal, as, at this point, polluters will gain no economic advantage from increasing the targeted environmentally damaging activity (see Figure 2.5). Where accurate calculation of marginal costs and benefits is not possible, the tax can be set on the basis of more qualitative environmental objectives such as the precautionary principle or another tenet of sustainable development (Ekins, 1999). In both cases, the tax is applied on each unit of pollution to create a continuous incentive for polluters to reduce or withdraw from the activity causing the externality effect (Stavins and Whitehead, 1992; Goddard, 1995).

Figure 2.5 Marginal costs and benefits of pollution using environmental taxes

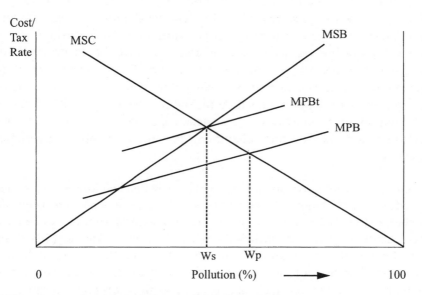

MSC private and social maginal costs of pollution reduction
MSB marginal socail benefits of pollution reduction
MPB marginal private benefit from reduced pollution
MPBt MPB after imposition of optimal tax per unit of pollution
Ws   Socially optimal level of pollution reduction
Wp   Level or pollution where damage from pollution equals private cost of pollution
     reduction

The aim of environmental taxes is to achieve 'socially optimum' levels of pollution abatement; that is, where the marginal social cost of further reductions in pollution units exceeds the marginal social benefit of that reduction. Initial reductions in pollution occur at relatively low cost (the actions that can be taken are straightforward) but, as greater reductions are pursued, higher investments and costs are incurred per abatement unit (MSC). Similarly, there is a lower *social* and *private* cost if lower pollution is achieved (MSB and MPB), the difference between MSB and MPB being an externality effect. The optimal pollution level for the *individual* is therefore, Wp, where MSC intersects MPB. Environmental taxes seek to increase the cost of pollution to the point where the marginal *private* benefit is the same as the marginal *social* cost and benefit (intersect MPBt, MSC and MSB), thus removing the dislocation between private and social benefits and costs.

Adapted from: Fenton and Hanley (1995: 1319)

Revenue-raising Taxes

In situations where an incentive tax or charge is not required or where it is not possible to calculate cost-covering charges accurately, economic instruments can still be used to raise revenue. This can be spent either on enhancing environmental services or to bring about a shift away from fiscal reliance on labour or product taxes, which are generally considered to be economically damaging (Gee and von Weizsäcker, 1994; Gee, 1997). When applied in a strategic manner, this is known as ecological tax reform and the ambition of such reform is to develop a taxation profile that balances social, economic and environmental prerogatives in a sustainable manner. With this form of environmental tax, revenue raising can become an explicit government objective.

Numerous alternative taxonomies to that provided by Ekins have also been identified. For example, Segerson (1996) classifies mechanisms according to whether they are *ex ante* – to prevent environmental damage – or *ex post* – to correct damage already inflicted – while Fenton and Hanley (1995) note that where product taxes are applied within the production, use and disposal cycle may alter their impact. Similarly, environmental taxes can be levied on the inputs to production, such as energy, so as to incorporate sustainability considerations across a range of products and services through a single policy instrument.

It is also clear that the main types of environmental tax are not mutually exclusive, as cost-covering and revenue-raising charges may produce incentive effects and *vice versa*. Ekins (1999) nonetheless argues that classifying environmental taxes in this manner clarifies their main objective, as not all objectives are entirely compatible. A cost-covering charge may not, for instance, be sufficiently high to induce an incentive effect, while a successful incentive tax may reduce revenues below those required for cost-covering or revenue-raising needs. Whilst such factors need to be considered in environmental-tax design, it is important also to appreciate that regulators' perceptions of objectives and the nature of incentives being created may vary from those of regulated companies.

Most economists acknowledge that environmental taxes should complement rather than supplant legislation but a key weakness of legislation in respect of environmental integration is that it only imposes threshold constraints on pollution. However, economic instruments monetise each unit of pollution, thereby creating a constant pressure for improvement (Pearce and Barbier, 2000). Taxation rates are also administratively simpler to adjust than legislative acts and, therefore, the cost of pollution can be altered to reflect changing economic and environmental circumstances (Goddard, 1995). Finally, because taxes are less prescriptive than legislation in specifying *how* firms manage environmental issues, sector-specific implementation methods can be adopted and, thus, compliance costs should be lower.

The behaviour-changing potential of economic instruments is not undisputed, however. Jacobs (1991) argues that whereas legislation compels firms or households to observe pre-defined environmental standards, incentives merely encourage them to do so. Ultimately individuals may choose to pay charges without checking pollution because of factors not considered within the economic model. More ardent critics even suggest that environmental taxes effectively create a licence to pollute since polluters and users may argue that they have paid their dues to society (Beder, 1996). Similarly, if environmental taxes are not a major cost to businesses in relation to other factors of production, they may be reluctant to invest heavily in pollution abatement technology (Jones, 1999). Either way, several iterations in taxation rates may be necessary to achieve the desired abatement level. Where highly price inelastic essential commodities are subjected to environmental taxes, the creation of an effective incentive tax may be politically unacceptable or economically damaging and, therefore, other forms of tax or policy instruments may be better suited to mitigating the environmental problem.

2.9   Tradable Permits

Tradable pollution or emission permits have their origins in the United States' Clean Air Act and the USA remains the only country to have applied tradable-permit systems on a large scale (Pearce and Barbier, 2000). The technique can operate in various ways but is based on the common principle that any increases in prescribed emissions or other polluting activities by one company must be offset by an equivalent or greater reduction elsewhere (Barde, 1997). Most tradable-permit schemes begin with government authorities setting a quantitative limit on the production or release of a pollutant they wish to control. This is then sub-divided into quotas, or pollution permits, which are issued or sold to businesses engaged in activities that create the prescribed pollutant. Permits can then be traded between firms, according to whether they are seeking to expand or contract the controlled activity.

The underlying ambition of trading is that companies with lower abatement costs will have a financial incentive to reduce pollution, as they are able to generate additional income by selling their excess permits. This revenue may also stimulate technological innovation in the industry. At the same time, companies that are less able to reduce their pollution because of the nature of their business, investment constraints or higher abatement costs have the opportunity to buy additional permits. Thus, the costs of achieving environmental standards is reduced through the use of market forces, since abatement is concentrated towards businesses with lower marginal pollution-reduction costs (Tietenberg, 1990). The creation and dissemination of new technologies financed by the trading of receipts may also make pollution abatement more affordable for a larger number of firms. Further environmental improvements can be achieved if the regulator chooses to reduce the number of permits in the trading system over time. This constricts the supply of permits and increases the value of those still in the system, magnifying the financial inducement for innovation and reduced pollution (Hahn, 1989).

As with economic instruments, there are also drawbacks to tradable permits. The most fundamental problem is where trading causes regional concentrations in degradation because optimum exchanges based on market logic – that of price – do not necessarily correspond with the equitable spatial or social distribution of pollution (Barde, 1997). It is therefore commonplace for regulators to introduce regional allocations of tradable permits, sometimes known as 'bubbles', to prevent acute localised pollution occurring as a result of permit trading. The environmental efficacy of the technique is also dependent on there being sufficient trading to create a financial incentive for firms to reduce their pollution (Cummings *et al.*, 2001). If all firms acquire permits then do nothing with them, pollution levels may remain more or less constant, though this can be counteracted by government authorities withdrawing or buying back permits. Despite these difficulties, there has been increased interest in tradable-permit schemes across the EU, particularly as a method for achieving its commitments under the Kyoto Protocol to reduce emissions of greenhouse gases (CEC, 2000b). Equally importantly in the context of this study, the UK government has introduced a tradable-permit scheme to help meet the requirements of the Packaging Waste Directive. This experiment with tradable recycling permits is reviewed in Chapters three and four.

2.10 Voluntary Agreements

As the limitations of command-and-control environmental regulation have been increasingly accepted, many governments have embraced the idea that environmental policy should involve closer co-operation between industry and public authorities in the definition of environmental standards and implementation methods (Börkey and Lévêque, 2000). This has led to an increase in the number of voluntary agreements either as a supplement to or a replacement for command-and-control regulation. Over 300 such schemes are now in operation in the member states, fuelled in part by the Commission's positive attitude towards the technique in the Fifth and Sixth EAPs (CEC, 1998a; 1999a; 2001a). Voluntary agreements have been applied in such diverse areas as: emissions from motor vehicles, industrial greenhouse-gas emissions, chemical industries and, importantly in the context of this study, packaging waste (Keay-Bright, 2000). As with economic instruments, voluntary agreements can be divided into three principal categories:

Unilateral Commitments
Unilateral commitments usually take the form of codes of conduct or guidelines developed by industries of their own volition (Börkey and Lévêque, 2000). The definition of environmental targets and monitoring provisions are determined by firms or sector associations, either independently or pursuant to a recognised voluntary environmental management system, such as ISO14000 or the European Eco-Management and Audit Scheme (EMAS). It is difficult to determine the precise extent of such voluntary agreements, as many companies have developed environmental policies that incorporate specific environmental commitments and might be considered to be unilateral commitments. However, they are not obliged to inform public administrators of these policies.

Public Voluntary Schemes
This type of scheme involves a group of firms agreeing to non-statutory performance, technology or management standards developed by public agencies. In contrast with unilateral commitments, the public body defines the pre-requisites of membership along with the standards to be applied and the monitoring arrangements. However, they remain voluntary in the sense that firms are not compelled to join the scheme. Key examples of public voluntary schemes in the EU are EMAS and ISO14000. Firms that have developed an environmental management system may register for EMAS or ISO accreditation. This provides them with an internationally recognised standard that may produce marketing and other financial benefits. Having been accredited, the firm's management system is subjected to periodic inspection and verification by independent auditors. In the case of EMAS, accreditation also requires disclosure of a public statement on the firm's environmental management policy. Another example is the European Eco-labelling Scheme, which was set up to provide a uniform system for determining the relative environmental impacts of particular categories of

consumer goods (Barnes and Barnes, 1999). In some cases, though not all, public voluntary schemes emerge where industry is faced with a choice between a voluntary instrument and another, more restrictive form of environmental regulation (Börkey and Lévêque, 2000). Voluntary agreements of this nature are therefore often concluded to pre-empt or avert government intervention.

Negotiated Agreements
Negotiated agreements are the most common form of voluntary agreement in the EU member states. They are formal contracts between public authorities and industry covering a specific aspect of environmental management and are established by means of joint negotiations. Such agreements may be either binding or non-binding depending on the constitutional rules of the state and the degree of liability exerted by the agreement. The majority of negotiated agreements in the EU are non-binding with the notable exception of The Netherlands, where 'covenants' established as part of the Dutch National Environmental Policy Plan are the main instrument of environmental policy (Öko Institut, 1998). Either joint or individual liabilities may be established by negotiated agreements. Under joint liability, all companies covered by the agreement are subject to penalties if the agreement fails. The normal sanction employed under joint responsibility would be the repeal of the negotiated agreement in favour of formal regulation. With individual liability, companies that evade their responsibilities (free ride the agreement) run the risk of prosecution. In practice, many negotiated agreements employ both forms of penalty. As Chapters three and four will argue, the voluntary agreements introduced for packaging waste in Germany and the UK were negotiated agreements designed to provide industries with greater latitude over how they met the regulatory standards set by the Packaging Waste Directive. The remainder of this discussion will therefore focus primarily on negotiated agreements.

The main reasons for the increased popularity of voluntary agreements in the EU include; reduced administrative costs for regulators and regulated, greater flexibility in how targets are achieved, increased innovation, rapid and relatively non-controversial implementation, and the removal of the need for legislation (Nunan, 1999). However, realising the environmental benefits of voluntary agreements requires a number of pre-conditions to be fulfilled. The first is that the agreement establishes environmental standards that are comparable to those that would have been secured using a legislative instrument, otherwise the negotiated agreement could lead to a deterioration in environmental quality (Segerson and Micelli, 1998). This risk is increased where industry representatives are more cognisant than public authorities of the technical issues surrounding the definition of 'optimal' environmental standards (there is asymmetry of information) (Segerson and Micelli, 1998; Nunan, 1999). This can lead to regulatory capture and the manipulation of agreements to the benefit of firms and the detriment of wider public interests (Böhmer-Christiansen, 1994).

The number of participants in an agreement also has a bearing on the success of negotiations (Bizer and Jülich, 1999). Where a larger number of firms or more diverse industry sectors are involved in negotiating an agreement, the range of stakeholder interests that must be satisfied increases and consensus is harder to reach. Finally, there is the problem of free riders. Even where agreements provide for sanctions against companies that ignore their obligations, firms may still be tempted to free ride the agreement if they feel they can get away with reaping the benefits of improved overall environmental standards without incurring the associated implementation costs. Free riding tends to increase where there is insufficient enforcement of the agreement or where individual liabilities are imprecisely defined. As enforcement procedures are never perfect, in practice there is invariably some 'leakage' between nominal and effective compliance (Heyes, 1998)[9]. However, Börkey and Lévêque (2000) claim that the inclusion of individual liability in Dutch environmental agreements has all but eliminated free riding.

2.11 Experiences with NEPIs in the European Union

The EU's experimentation with NEPIs reveals a number of interesting patterns, some of which are consonant with the customary dynamics of European environmental policy and others that are more surprising. A report by the European Environment Agency (EEA, 2000) shows an upsurge in the number of new policy instruments deployed by the member states, particularly environmental taxes and voluntary agreements. Several 'leader' states – Denmark, The Netherlands and Finland – have developed numerous environmental taxes, while The Netherlands is unique in making voluntary agreements the mainstay of its environmental policy. It is also interesting to note that in 1997 a number of more environmentally cautious states (the UK, Greece, Ireland and Portugal) raised a greater proportion of their total tax revenue from environmental taxes than did Germany, the EU's main environmental agitator. Although Germany has introduced a far-reaching ecological tax reform since the EEA report was compiled, wide variations nonetheless remain in the deployment patterns of environmental taxes in the member states (Table 2.6).

Table 2.6   Environmental taxes and charges in EU member states

Tax on	A	B	DK	FI	FR	GE	GR	IR	IT	L	NL	P	SP	SW	UK
Energy	░	░	░	░	░	░	░		░			░	░		
CO2			░		█	█		█	░		░				
Transport	░		░	░	░		░	░	░		░		░		
Diff. car tax				░		█									█
S in car fuel			░												
Water effluents	░	░	░		░	░			░	░	░				
Waste end	█			█	█		░	█	░					█	█
Dangerous waste			░												
Tyres	░		█												
Disposable razors		░													
Beverage containers		░												░	
Disposable cameras		░													
Raw materials				░											
Packaging	█	█	█	█	░	░			█	█	░		█	█	█
Bags			░							░					
Disposable tableware			░												
Pesticides		█	░	░											
CFCs	░	░													
Batteries	█	░	░						█						
Light bulbs			░												
PVC			█												
Lubrication oil			░	░					░						
Fertilisers														░	
Paper/board	░	█	░			░									
PE											░				
Solvents			█												
Aviation noise											░				
NOX					░									░	
SO2					░				░						
Minerals			░	░							░				

░ In existence in 1996   █ Introduced 1997–2000

Source: EEA (2000: 26, adapted)

In many cases, however, national governments have been fairly tentative in their experimentation with NEPIs, preferring instead to develop repertoires of policy instruments that combine NEPIs with more traditional forms of regulation. This has enabled governments to retain the reassurance of having established clear performance standards whilst allowing them to utilise NEPIs to resolve implementation problems, such as the financing of new environmental services. NEPIs have also often been deployed or adjusted iteratively to avoid shock therapy and to take account of changing market circumstances. The UK Landfill Tax, for instance, was introduced in 1996 at £7 per tonne for active waste but has been increased in stages to raise the incentives for industry to recover and recycle waste (Porter, 1998).

By contrast, relatively few NEPIs have been introduced directly by the EU because of the legal constraints on its intervention in national affairs. Some less controversial NEPIs have been introduced as European directives and regulations, including the EMAS and Eco-label regulations, though take up of these has been lower than for equivalent national systems. However, the Commission's attempts during the 1980s and 1990s to introduce an EU carbon tax is a prime example of the problems involved in harmonising NEPIs across the EU. Despite the obvious trans-national dimension of the climate change issue and, consequently, the apparent justification for EU action under the subsidiarity principle, some member states feared that carbon taxes would damage their national industries. Germany, for example, expressed concerns about the impact of carbon taxes on employment in its coal industry even though it supported the principle of environmental taxation (Zito, 2000). Britain's objections were more politically motivated and questioned the democratic legitimacy of transferring taxation powers from the national to the European level. As any EU measures to change taxation policy in the member states must receive unanimous support from the member states, Britain was able to invoke Article 93 of the Treaty to veto the Commission's proposals despite repeated attempts at mediation by other member states (Barnes and Barnes, 1999)[10].

In spite of this policy impasse, the EU debate on carbon taxes and international discussions at the Kyoto Summit convinced several member states that the most effective way to tackle climate change was to implement national energy taxes. In 1999 the German government introduced climate protection measures as part of its Ecological Tax Reform, in 2001 the UK introduced the Climate Change Levy, and other member states such as Belgium and Sweden have either instigated or expanded national carbon tax schemes (HM Treasury, 1999; *BMU*, 2000). Although a European Climate Change Strategy is still on the Commission's agenda, national carbon taxes have the critical advantage of enabling member states to fashion policies to suit national circumstances and their varying targets under the Kyoto Protocol (CEC, 2000b). The UK, for example, agreed an 80 per cent reduction in the Climate Change Levy for energy-intensive sectors in order to soften the impact for industry. In exchange, these sectors have signed voluntary agreements for emissions reductions, which they must uphold in order to retain their rebate. Germany has introduced a similar scheme but again tailored to the circumstances of the German economy. The carbon tax proposal therefore demonstrates both the limitations and the potential contribution of the EU towards the dissemination of NEPIs. On the one hand, subsidiarity arguments have severely curtailed the Commission's options for accelerating and harmonising the deployment of NEPIs in the member states. On the other hand, the Commission's role as policy advocate has prompted greater discussion on use of flexible regulation in environmental policy.

It is nonetheless the case that patterns of NEPI deployment across the EU have been overwhelmingly nationally led, again reflecting the dualism between policy formulation and implementation that exists in the EU. The Commission must therefore be mindful that uneven patterns of NEPI deployment in the member states may become a major source of technical trade barriers in the internal market. Because of their fiscal nature, environmental taxes and tradable permits require especially close scrutiny to reassure industry that they are non-discriminatory. The EU's rather unwieldy infringement procedures may therefore be tested to the full if the recent upsurge in state-led NEPIs continues unabated.

2.12 Conclusions

NEPIs have become increasingly popular with European and national policy-makers because they offer the possibility of improved environmental protection at lower cost than traditional regulatory techniques. Environmental and ecological economists in particular have exerted a seminal influence through their exploration of the inter-relationships between market behaviour and environmental degradation. The current popularity of NEPIs has also been prompted by less normative factors, including a greater receptiveness amongst policy-makers towards the logic of the market, the influence of cognate epistemic communities, and international pressures from the World Trade Organisation (WTO) and the OECD.

As NEPIs have moved from their cognitive laboratories into the policy armouries of the member states, several, not entirely unanticipated, complexities have emerged. First, it has become apparent that NEPIs need to be tailored meet a range of policy priorities. Sustainable development implies the melding not just of environmental and economic objectives but also of social criteria. The distributional impacts of environmental taxes have therefore become a prominent issue, as has the influence of indirect market factors on the performance of NEPIs. This has led to major experimentation and variation in the design and deployment of NEPIs. Whilst it is too soon to assess the effectiveness of NEPIs that have only recently been implemented, some economists are already lamenting that charges and marketable permits are rarely introduced in their textbook form but are instead grafted, sometimes inappropriately, onto existing regulatory systems (Goddard, 1995). In such cases, it is argued, the under-performance of NEPIs may stem from regulatory failure rather than intrinsic defects in economic theory.

The deployment of NEPIs also raises a number of intriguing political issues. The first is whether such instruments are actually new and innovative (Jordan *et al.*, forthcoming). Many of the techniques are only novel in respect of their application in environmental policy, as they have existed in various guises in other policy areas for many years. Equally, how far do they represent part of a fundamental learning process or paradigm shift in environmental policy? Alternatively, are NEPIs being deployed in a fairly *ad hoc* fashion by governments desperate to try any new idea that might resolve longstanding environmental problems? The broad spectrum of areas where NEPIs have (and have not) been introduced in the member states as well as disparities in the type and design of instruments introduced would suggest that much experimentation is taking place but that a clear strategy has yet to emerge. Finally, it is apparent that the EU has only limited capacity to cohere NEPI patterns across Europe, since its powers to challenge the way in which member states manage the implementation of environmental policies are restricted. Whilst these constraints have historically helped to maintain the flexibility and democracy of the EU environmental programme, maintaining the cohesion of environmental policy *vis-à-vis* other areas of EU activity, particularly the internal market, may prove increasingly difficult if nationally-led NEPIs continue to grow in popularity.

Notes

[1] The term Council of Ministers is used throughout this book, as this was the Council's official title at the time the Packaging Waste Directive was negotiated. The Council of the European Union should not be confused with the European Council, the term used to describe meetings of EU heads of state.

[2] Using the example of the EU's policy on hazardous waste, Zito (2000) argues that scientific

knowledge of environmental risks is rarely used by the Commission as the primary basis for proposing new environmental standards, even in highly technical policy areas. He claims that efforts to depoliticise EU decision-making through the use of science are frequently thwarted, especially when important economic interests are involved. Similar conclusions may be drawn concerning the failure of the Common Fisheries Policy (CFP) to conserve stocks of North Sea cod and hake (Payne, 2000; CEC, 2001b).

[3] Due to space constraints, detailed reviews of other perspectives on European integration, such as neo-functionalism and consociationalism, are not provided here. For fuller accounts, see Slater (1982), Moravcsik (1991) and H. Wallace (1996a).

[4] Though Sbragia (1996) also notes that environmental ministers enjoy higher prestige within the Council than they do domestically and can therefore add credence to their positions by demonstrating unanimous support for substantive policy advances.

[5] Weale (1996) cites the Integrated Pollution Prevention and Control Directive (96/61/EC) as an example of EU policy upgrading national legislation. The UK attempted to have its national legislation adopted as EU law but the final directive became far broader in scope than the UK originally intended.

[6] Named after Article 226 of the Nice Treaty, which covers enforcement procedures.

[7] These actions are for failure to implement the Bathing Water and Environmental Impact Assessment directives respectively and follow initial Court of Justice rulings against the two member states.

[8] Examples include the installation of air scrubbers on incinerator chimneys and the introduction of lead-free fuel for private petrol vehicles.

[9] The UK Environment Agency estimates that the average business compliance rate with existing legislation is approximately 74% (Heyes, 1998).

[10] Article 93 of the Treaty requires the Council to act unanimously on the harmonisation of taxes and only allows EU intervention where it is considered necessary to ensure the establishment and functioning of the Single Market. Britain argued that carbon taxes could be dealt with more selectively and sensitively by individual member states and, therefore, EU action was not appropriate under the principle of subsidiarity.

Chapter 3

The Packaging Waste Directive

3.1   Introduction

The previous chapter contrasted normative theories used to inform the selection of environmental policy instruments with the political realities of environmental decision-making in the European Union. It was suggested that although economists have attempted to de-politicise environmental problems through the propagation and dissemination of 'scientific' tools and techniques, political considerations pervade all aspects of policy, including the formation of legislation and the design and deployment of implementing mechanisms. Successive EU Treaties have also altered the status of the EU programme and the roles of the respective Community institutions in environmental decision-making. It is therefore apparent that the practical implementation of EU environmental directives cannot be properly understood without first appreciating the political and institutional contexts in which policies are formulated (Demmke, 1994; Archer and Butler, 1996; Segerson, 1996).

Thus far the debate has analysed EU environmental policy in general terms. The purpose of this chapter is to examine the dynamics that accompanied the negotiation and implementation of the Packaging Waste Directive. The discussion begins by analysing the pressures that led to the Directive, where it is argued that an accretion of member-state actions and legal precedents made the existence of unilateral national laws on packaging waste increasingly untenable. Despite a general consensus that harmonising legislation was required, negotiations on the Directive led to intense interest-led bargaining between environmental leader and laggard states. The next section traces these arguments and the way in which they were mediated through the EU decision-making machinery. It is argued that the need to appease the more environmentally cautious states – a pre-requisite for gaining majority support for the Directive – caused many aspects of the policy to be diluted. At the same time, the environmental leader states insisted on flexibility clauses to ensure their environmental preferences were catered for. The following section traces the practical implementation of the Directive in the United Kingdom and Germany. Specifically, it describes the implementing measures employed in each country and assesses the extent to which command-and-control and flexible regulation have been employed. The use of NEPIs in Britain and Germany is then compared in terms of their orientation towards environmental stringency and economic efficiency, and concluding remarks are offered.

The main argument presented in this chapter is that state control over the implementation of European environmental policy exacerbates the uneven national standards created by the EU's negotiating procedures. Whilst state-led implementation is an essential pre-condition for accommodating issues of sovereignty and the environmental capabilities of each member state, the implications of this for sustainable development at a strategic level should not be overlooked. Evidence from the Packaging Waste Directive suggests that the uneven application of EU policies has been further increased with the move towards NEPIs, as these instruments can be finely honed to reflect particular national policy agendas. Within this climate, balancing the efficiencies that NEPIs offer with the maintenance of a coherent EU environmental programme is an especially tricky task.

3.2   National Policies and the Packaging Waste Directive

The Packaging and Packaging Waste Directive (94/62/EC) was formally agreed by the Council of Ministers and the European Parliament on the 20th December 1994. Although the Directive formalised standards for the management of packaging waste throughout the Union, the initial impetus for the policy did not come from the EU but the actions of several environmental leader states. The negotiation of the Packaging Waste Directive is therefore a prime example of the 'push-pull' dynamic in European environmental policy-making (Sbragia, 1996).

The first moves towards European regulation on packaging waste came in 1982 when the Danish government introduced legislation requiring that all beer and soft drinks placed on its markets should be sold in refillable containers that carried a refundable deposit. This measure effectively banned the use of metal cans and plastic bottles for these products because they are not generally suitable for re-use, though they can be recycled. The Danish law also stipulated that all beverage containers must receive prior approval from the Danish authorities (Barnes and Barnes, 1999). This generated an obvious tension between the Danish law and the free movement of goods in the Single Market, as it created conditions whereby non-Danish beverage producers incurred higher transport costs than domestic manufacturers in order to establish the necessary return and refill systems (Bailey, 1999a). The Commission's initial response to complaints from other member states that the Danish law was tantamount to protectionism was to propose the so-called Beverage Cans Directive of 1985 to harmonise national laws along the lines of the Danish approach. However, this was rejected by the Council and, as criticism of the Danish legislation intensified, in 1988 the Commission launched the 'Danish Bottles Case' (302/86, ECR 4607) in an effort to have the law repealed (Porter, 1998). In one of the most significant rulings in EU environmental case law, the Court of Justice upheld the environmental objectives of the Danish legislation on the grounds that environmental protection was a core policy of the EU. However, it also declared that this precedent would only apply where there was not already satisfactory EU legislation and where national measures were proportionate with the scale of the problem being addressed (Lister, 1996).

The issue of free trade and waste surfaced again in 1992 when the Belgian province of Wallonia banned waste imports from other EU states and other regions of Belgium. The embargo followed a sharp increase in international waste shipments since the late 1980s, particularly from Germany. The legal issue at stake was whether waste was a tradable commodity that was permitted free movement across the EU under Articles 30 and 36 of the Treaty of Rome or whether waste management should be treated as a purely environmental matter. It is noteworthy that the EU had already created a system under Directive 84/631/EEC for controlling the trans-frontier shipments of wastes, which prohibited blanket bans and only entitled national authorities to reject consignments where important public policies, such as health, were threatened (Lister, 1996). This meant that the Wallonian authorities could not argue that unsatisfactory EU controls were in place, as the Danish government had for beverage containers. However, the ECJ declined to forbid the Belgian law (Case C-2/90, *Commission vs. Belgium* [1992] 1 ECR 4431), deciding that Article 130r of the SEA specifying that damage to the environment should be remedied at its source, applied in this instance. This interpretation of the proximity principle meant that, wherever possible, the international movement of waste should be kept to a minimum. The Court also agreed with the Belgian position that unrestricted waste shipments to a small region constituted a severe threat to local environmental quality and warranted more extreme measures. This ruling therefore established the more general principle that where there was a particularly acute environmental problem, 'exceptional' exemptions to the free movement of goods could be granted (Lister, 1996). At the same time, the case again highlighted the difficulties arising when member states introduce national environmental laws that adversely affect free trade in the internal market (von Wilmowsky, 1993).

But it was the German *Verpackungsverordnung* (Packaging Ordinance) of 1991 that provided the final momentum for the introduction of the Packaging Waste Directive (London and Llamas, 1994). The Ordinance first established strict recycling laws, which were not theoretically problematic for EU free trade because they primarily targeted post-consumer waste. However, it also required manufacturers and retailers to take back packaging waste from consumers for re-use or recycling and included provisions for a mandatory deposit-refund system for beverage containers if the Ordinance's 72 per cent re-use target for cans and bottles was not met in any given year (*Umweltbundessamt, 1991*). As with the Danish can ban, these latter requirements potentially restricted free trade in the Single Market because they imposed additional transport and administrative costs on non-German drinks manufacturers (Waite, 1995; Bailey, 1999a). The deposit-refund system was in fact deferred for German manufacturers and distributors that participated in an industry-led system of recovery and recycling for packaging waste, the details of which are discussed later.

The Commission nonetheless felt that elements of the Packaging Ordinance posed a serious threat to free trade because the German economy is the largest in the EU. At the same time, it was concerned about the implications of challenging national legislation that promoted a precautionary approach to waste management similar to that advocated in the EAPs. The move to prosecute Germany was also opposed by the Commission's own Environment Directorate (then DGXI), including Ludwig Krämer, then head of the Waste Policy Unit and a German environmental lawyer renowned for his green views and disinclination to compromise on matters of legal principle (Jordan, 1999). Notwithstanding this, the Commission remained under pressure from several member states, most notably Britain, who felt the trade restriction was unacceptable and, in 1992, it began proceedings to have the deposit-refund clauses of the Packaging Ordinance repealed.

Initial exchanges took the form of bilateral discussions between the Commission and the German federal government. In addition to arguing that the extra costs of the take-back provisions and the general preference for re-use over recycling restricted the free movement of goods, the Commission claimed that Germany's high recycling targets could not be justified in terms of environmental protection since some of the waste was merely being transported to other member states for landfill (Perchards, 1998). The German government was under considerable pressure not to relent on its policies, however. The upper house, the *Bundesrat*, which represents the views of the German states (*Länder*), supported the take-back and deposit schemes and used the fact that the *Länder* are responsible for managing most aspects of waste policy to press their case (Haverland, 1999). The German government used three arguments in its defence. First, it noted that beverage containers represented only a small proportion of the total waste generated in Europe and therefore argued that the potential trade restriction was minimal. Second, it produced life cycle assessments by the Fraunhofer Institute in Munich, which indicated that the re-use of packaging waste was generally less environmentally harmful than recycling (see also Beynon, 1993). Consequently, the German government felt that the Ordinance was proportionate and upheld the EU's own ambitions with regard to sustainable waste management expressed in the Fifth EAP (CEC, 1992). Although the EAPs do not contain the same legal weight as directives or regulations, the Commission could not disregard this point lightly considering the precedents set by the Danish and Wallonian judgements. Finally, the German government agreed with the Commission that the general market quotas for refillable packaging should ultimately be replaced by product-specific quotas but maintained that it would be counter-productive to dismantle these elements of the Ordinance in the interim (Perchards, 1998).

One of the more intriguing features of these discussions was the fact they took place at a time when the Commission had already tabled proposals for an EU directive on packaging waste, though these did not seek to replicate Germany's mandatory deposit-refund scheme in EU law. Whilst the pressure for proceedings against Germany had begun in 1992, a framework agenda for Europeanising national laws had been agreed by the time detailed discussions took place. This demonstrates not only the cumbersome nature of EU enforcement procedures, which can take several years to conclude, but again highlights the prominence of free-trade arguments in determining the acceptability of individual national laws designed to protect the environment (Collins and Earnshaw, 1993).

Throughout 1992 and 1993 the pressure for an EU directive mounted as more states considered the benefits of national laws as a means of eliminating trade restrictions whilst at the same time addressing their own packaging waste problems. France in particular gravitated towards this opinion and sought to pre-empt any move towards an EU directive by introducing its own *Décret du 01/04/92 déchets d'emballages menagers* 92-377 (Ordinance on used packaging from households) (London and Llamas, 1994). The French Packaging Ordinance borrowed extensively from the German legislation in terms of recovery, recycling and monitoring systems but did not include specific targets except for its main recycling organisation, *Eco-Emballages* (Hagengut, 1997). The Commission was therefore faced with the prospect of a profusion of disparate national laws controlling packaging waste. This persuaded Commissioners that an EU directive was the most logical way of avoiding serious trade restrictions whilst simultaneously re-invigorating its own policies on packaging waste, which had stalled with the rejection of the Beverage Cans Directive (Haverland, 2000a). Fortuitously for the Commission, the most vociferous opponent of Germany's re-use quotas, Britain, also supported this move on the grounds that it maintained a level playing field for European competition.

Unlike some new EU policies, therefore, the question of subsidiarity was not a major stumbling block for the introduction of the Packaging Waste Directive as, for various reasons, the majority of major member states supported the initiative. In short, Germany succeeded in pushing its domestic policies onto the EU agenda because governments like the UK, which might have sought to block the initiative, generally accepted the environmental and trade benefits of a European policy designed to promote recycling. However, the question remained as to whether the Directive should be introduced under the environmental requirements of Article 130r or Article 100a, which covers the completion of the common market. Because the main sticking point between national and EU provisions was the potential constraint on the free trade in packaged products caused by re-use quotas and uneven recycling targets, it was agreed that the Packaging Waste Directive should be expedited under Article 100a (Golub, 1996)[1]. This decision had its own implications, however, as it failed to diffuse the dispute between the Commission and Germany on take-back quotas and deposit-refund schemes. The first hurdle had nonetheless been successfully negotiated and the member states prepared to debate the Packaging Waste Directive.

3.3   Negotiating and Transposing the Packaging Waste Directive

In line with standard EU policy-making procedures, the Commission drafted initial proposals for the Packaging Waste Directive for discussion by the Council of Ministers and European Parliament. The pre-draft version was co-ordinated by DGXI and reflected its environmental remit. The main elements of the strategy included; a per-capita limit on the consumption of packaging waste of 150kg/year, mandatory recovery and recycling rates of 60 per cent and 40 per cent within five years – rising to 90 per cent and 60 per cent in ten years – and a hierarchy of disposal options that prioritised, in the following order: source reduction (prevention), re-use, recycling, incineration and, finally, landfill (Golub, 1996). These suggestions were opposed by other Directorates and industry groups, however, as they felt that overly ambitious targets failed to reflect the underdeveloped state of European recycling markets

(Haverland, 1999). In order to gain the unanimous approval of the College of Commissioners, the final negotiating text dispensed with per capita waste quotas, five-year targets and the hierarchy of disposal options.

The proposal then received its first reading by the European Parliament where, under the co-decision procedure applied to Article 100a directives, amendments could be suggested. The Parliament's Environment Committee, which produces reports of recommendations to the chamber on environmental matters, proposed the reinstatement of the lapsed measures with the exception of the per capita limits (European Parliament, 1993). However, the Commission was again pressed to remove any binding reference to the waste hierarchy and also sought to restrict attempts by environmental leader and laggard states to include opt-up or derogation clauses allowing them to pursue potentially disparate national agendas with impunity. The protection of free trade was therefore again paramount in the Commission's deliberations.

These divisions re-emerged forcibly when the Commission's revised proposal went before the Council of Ministers. The Council was effectively split into two coalitions, an environmental-leader group consisting of Germany, Denmark and the Netherlands, and a more sceptical faction headed by Britain, Ireland, Spain, Portugal, Greece and more tacitly supported by France and Italy. The latter coalition consisted of member states concerned about the economic impact of these proposals and the scientific justification for introducing such high recycling targets. The size of this faction was sufficiently large that the German-led coalition was unable to secure a qualified majority in Council and, in order to gain support for any measures at all, many of the more demanding elements of the Commission's proposal were dropped or severely diluted. The key casualties in these negotiations were the commitment to ten-year targets, the restrictions on derogations, and harmonised recovery and recycling targets, which were replaced by target 'bands.' Although the European Parliament attempted to reintroduce some of these provisions in its second reading and a conciliation committee was convened to resolve the wording on the use of economic instruments, the majority Council grouping succeeded in significantly weakening the scale and scope of the Directive's requirements (European Parliament, 1994). A summary of proposals at each stage of the negotiations is shown in Table 3.1.

The final text of the Directive nonetheless establishes legally binding targets for the recovery and recycling of packaging waste across the EU. By the 30th June 2001, each member state must introduce systems which ensure that 50-65 per cent of the packaging sold on its markets is recovered and that 25-45 per cent is recycled, with a minimum of 15 per cent recycling for each packaging material (OJEC, 1994a). When calculating the amount of packaging

Table 3.1   Development of the Packaging Waste Directive

	Comm1	Comm2	EPEC1	EP1	Comm 3	Common Position	EPEC2	EP2	Adopted
Per capita limits	Yes	No	No	No	No	No	No	No	No
Minimum use of recycled materials	No	No	Yes	Yes	No	No	No	No	No
Hierarchy of preferred disposal	Yes	No	Yes	Yes	No	No	Yes	No	No
Opt-ups	No	No	Very broad	Limited	Limited	Very limited	Very limited	Very limited	Very limited
Derogations	No	No	Very limited	Very limited	Very limited	Yes	Yes	Yes	Yes
Five year targets									
• total recovery rate	60%	No	60%	60%	60%	50-65%	50%	50-65%	50-65%
• total recycling rate						25-45%	25%	25-45%	25-45%
• recycling rate per material	40%	No	40%	40%	40%	15%	25%	15%	15%
Ten year targets									
• total recovery rate	90%	90%	90%	90%	90%	No	No	No	No
• recycling rate per material	60%	60%	60%	60%	60%	No	No	No	No
• maximum landfill & incineration	10%	10%	10%	10%	10%	No	No	No	No
• heavy metals ban	No	No	Yes	No	No	No	No	No	No

Notes: Comm 1   =   Environment Directorate's pre-draft objectives
　　　　 Comm 2   =   Commission draft directive (12.10.92)
　　　　 EPEC1   =   First report by the Environmental Committe of the European Parliament (8.6.93)
　　　　 Comm 3   =   Revised Commission proposal (9.9.93)
　　　　 EPEC2   =   Second report by the Environment Committee of the European Parliament (7.4.94)
　　　　 EP2   =   Second reading by the European Parliament (4.5.94)

Source: Golub (1996: 323) published by Blackwell Publishing

it is required to recover and recycle, each member state is obliged to include all packaging produced and imported for domestic consumption but may exclude packaging produced for export. Whilst there is no reference to a prescriptive waste management hierarchy, the Directive specifies that these systems should 'as a first priority' promote the reduction of packaging waste and also encourage the design of packaging so as to facilitate its re-use and recycling[2]. Additionally, national systems should assist in the development of end markets for recyclates. However, no mandatory standards were set for these objectives, despite them being termed 'Essential Requirements' in the Directive. Article 11 outlines a timetable for reducing the concentrations of lead, cadmium, mercury, and hexavalent chromium present in packaging.

The Directive also contains a series of derogations and opt-ups despite the Commission's concerns over free trade. Article 6 permits member states to go beyond the targets established by the Directive on the condition that these measures avoid distortions of the internal market and do not hinder other states' compliance with the Directive. At the other end of the spectrum, Greece, Ireland and Portugal, by virtue of their large number of islands, rural and mountain areas and low levels of packaging consumption are only required to recover 25% of their packaging by 2001 and may delay full compliance until 2005 (OJEC, 1994a: 14). So, whilst the Packaging Directive was originally intended to align national recycling laws, the negotiation of the policy at the EU led to the creation of a broad basket of requirements designed to appease all the main parties. The dominance of the laggard coalition was therefore not absolute, as the minority Council grouping was able to extract concessions in the form of opt-ups that reduced the need to alter their existing policies. Although these compromises were ultimately necessary to maintain an accord in the EU process, one frustrated Parliamentary deputy branded the final Directive a 'mess of ill-assorted, inconsistent compromises' (European Parliament, 1994: 12). This strategy nevertheless permitted flexible implementation under the banner of 'approximated' EU law. Finally, on the subject of implementing mechanisms, the preamble to the Directive refers to voluntary agreements and Article 15 paves the way for the use of economic instruments by setting out the environmental principles and Treaty conditions they must observe. Within this framework, decisions as to the precise format of economic and other policy instruments are left largely to the discretion of the individual member state.

Having reached agreement in the Council, member states were then required to transpose the Directive into national law by 30th June 1996 and begin the process of developing implementing mechanisms. This created varying degrees of adaptation pressure. Those states with existing legislation, namely

Germany, France, Denmark and the Netherlands, had secured the concessions they thought necessary to avoid major changes to national legislation, though none of their measures was entirely compatible with the Directive (Hayes-Renshaw and Wallace, 1995). Britain and the Mediterranean states faced a sterner task despite the erosion of the Directive's standards, because they possessed less well-developed recycling infrastructure.

In order to ensure that EU and national laws were comprehensively aligned, the Commission pursued early transposition failures with considerable vigour. In 1998, no fewer than six member states received reasoned opinions for incorrect transposition. Britain, Belgium, Ireland and Portugal were all reprimanded for not transposing the Directive's 'Essential Requirements' (the commitment to waste prevention and re-use), whilst Greece and Luxembourg were cautioned for failing to adopt legislation by the required deadline (ENDS, 1997a; VALPAK, 1998, CEC, 1998b). In response, in 1998 the British government introduced its own Packaging (Essential Requirements) Regulations (Department of the Environment, Transport and the Regions (DETR), 1998a). The Commission was equally intolerant of non-agreed exclusions from the Directive. Britain and Finland were cited over their failure to transpose the Directive in Northern Ireland and Aaland respectively (CEC, 1998c). Although both regions have special status under the EU treaties (the reasons cited for the delays), Britain subsequently introduced legislation for Northern Ireland but an application was lodged with the Court of Justice to begin non-compliance proceedings against Finland (ENDS, 1997a).

Two outstanding cases remain at the time of writing. These relate to renewed attempts by the Commission to overturn Germany's re-use quotas and the Danish ban on non-refillable beverage containers. The German case is reviewed in detail in the discussion of the German implementation of the Directive. Aside from these cases, incomplete transposition of the Directive has generally been limited to states that have experienced genuine problems gaining domestic agreements on the design of implementing mechanisms and the apportionment of legal responsibilities. In the case of Luxembourg and Ireland, this has been further complicated by the fact both countries export most of their recyclable waste and, thus, need to develop secondary reprocessing agreements. Notwithstanding the German and Danish cases and the in-built flexibility of the Directive, which itself encourages considerable policy diversity in the member states, the Commission's management of the transposition process has ensured that formal implementation has not been the source of major variances between national policies and EU requirements.

3.4 National Packaging Waste Management Systems

The Commission and Council of Ministers sought to resolve political difficulties surrounding the negotiation of the Packaging Waste Directive by incorporating measures that catered for the implementation preferences and capabilities of each member state. The need to secure a Council majority within a procedure and political climate that demanded consensual bargaining therefore led to a directive which effectively sanctioned persistent diversity in member-state packaging laws because it allowed each state considerable latitude in determining the objectives it wished to pursue. Simply put, the Directive's banded targets, derogations and opt-ups meant that two member states could achieve markedly different policy outcomes and both still claim to have faithfully complied with EU law. The emphasis was therefore very much on flexible implementation.

The member states nonetheless faced two common challenges in implementing the Directive. First, most states needed to expand national recycling infrastructure so as to collect and reprocess the amounts of packaging required by the Directive. In the United Kingdom, for example, reprocessing capacity was required to double from 1996 levels to meet EU standards (Bailey, 1999b). Financial mechanisms were therefore needed to fund the necessary investments which, according to the Directive, should be levied in accordance with the polluter pays principle (Article 15). This meant either a consumer tax, using the interpretation that the general public creates waste by demanding packaged products, or recycling charges for producers. Second, national regulations needed to ensure that the Directive's targets were met without impeding EU trade. In addition, the majority of member states were keen to achieve these objectives without causing excessive damage to the competitiveness of their domestic industries. The most logical way of achieving this, in the opinion of most governments, was through negotiated agreements with industry on the design and deployment of implementing mechanisms. Across Europe, governments took the view that legislation alone would not achieve these multiple objectives and therefore sought to apply more flexible and dynamic methods of regulation. In most cases, this has led to the adoption of 'voluntary' agreements and environmental charge or tax schemes.

Four basic models of compliance have been identified by Bailey (1999a) and Haverland, (1999, 2000a, 2000b); the UK's Producer Responsibility Obligations (Packaging Waste) Regulations 1997 (the Regulations), Germany's command-and-control *Verpackungsverordnung*, voluntary agreements between industry and government under the Dutch Packaging Covenant, and

Table 3.2 National measures to implement the Packaging Waste Directive

Member State	2001 target (% of packaging weight)			Packaging Re-use Provisions	Packaging Refill Quotas	Tax on single trip containers	Landfill tax	Producer Responsibility
	Reduction	Recycling	Recovery					
Austria		10 - 70 [a]	80	Some	Yes	No	n/a	Yes
Belgium		50	80	Yes	Yes	Some	n/a	Yes
Denmark		25 - 45	50 - 65	Yes	Yes	Yes	Yes	Yes
Finland	6	42	61	Yes	No	Yes	Yes	Yes
France		25 - 45	50 - 65	No	No	No	Yes	Yes
Germany		45	65	Yes	Yes	Yes	n/a	Yes
Greece		-	30	No	No	No	-	-
Ireland		25	33	Some	No	No	Yes	Yes
Italy		25 - 45	50 - 65	No	No	No	Yes	Yes
Luxembourg		45	55	Yes	Yes	Yes	n/a	Yes
Netherlands	10	45 [b]	65	Yes	Yes	No	Proposed ban on packaging	Yes
Portugal	10	-	25	Yes	Yes	No	n/a	Yes
Spain		25 - 45	45 - 65	Some	No	Some	n/a	Yes
Sweden		30 - 90 [a]	70	Some	No	Repealed 1993	n/a	Yes
UK		18 [c]	56	Very limited	No	-	Yes	Yes

[a] Separate targets set for each packaging material covered by national legislation.
[b] The second Dutch Packaging Covenant sets an overall recycling target of 65% for 2001. This is a voluntary agreement, however, not part of binding legislation.
[c] The UK recycling target is a minimum rate for each material covered by the Producer Responsibility Regulations.

Source: Bailey (1999a: 556) published by Blackwell Publishing

the Danish system of integrated waste management. Because of space restrictions, this study concentrates on the British and German models. A summary of national measures is provided in Table 3.2 but for a fuller account, see Hagengut (1997) and Perchards (1998). The following sections describe the structure of the British and German implementation strategies and investigate the use of NEPIs and 'command-and-control' legislation to implement the Packaging Waste Directive.

3.4.1 The German Model

The German Packaging Ordinance of 1991 has been described as 'the most prescriptive and demanding piece of environmental legislation passed by any European government with regard to packaging waste' (Waite, 1995: 137). It makes manufacturers and distributors responsible for the recovery and recycling of their packaging waste outside the public waste disposal system by requiring them either to remove secondary packaging from goods before offering them for sale or provide in-store facilities where consumers can leave used packaging (Raymond Communications, 1998). The Ordinance also establishes stringent re-use targets, notably the 72 per cent refill quotas for beverage containers, and provisions for the automatic introduction of a deposit-refund system should this not be achieved in any given year. This ruling applies in each German state and permits individual *Länder* governments to impose mandatory deposit systems in their territories if re-use quotas are not met (Michaelis, 1995).

The federal government's emphasis on strong command-and-control environmental regulation is also reflected in the Ordinance's interpretation of the waste management hierarchy, where the reduction, re-use and recycling of packaging waste were made key policies objectives but incineration with energy recovery was largely frowned upon as a method of waste recovery[3]. Interpreting the Ordinance as an entirely command-and-control policy is somewhat misleading, however, as numerous NEPIs have been deployed by the German authorities to expedite their policies. The first of these is a negotiated agreement between industry and government waiving mandatory take-back and deposit-refund schemes for manufacturers and distributors taking part in an industry-organised system for collecting, sorting and recycling used packaging, the *Duales System Deutschland* (Dual System or DSD). This concession only applies to sales packaging, however, and the recovery and recycling of secondary and transport packaging must be organised independently by obligated businesses (Michaelis, 1995). In fact, negotiations for the formation of the DSD were concluded before the Ordinance was adopted though, in truth, industry was offered few alternatives except the immediate enactment of the Ordinance's take-back and deposit-refund provisions (Haverland, 1999). Command-and-control was therefore used partly as a threat by the federal government to 'persuade' industry to co-operate with its agenda. 95 companies from the retail, consumer goods and packaging sectors originally formed the DSD. By 1998, the number of shareholders had increased to 600 (Whiston and Glachant, 1996) and over 18,000 businesses now use DSD's systems to discharge their recycling obligations (DSD, 1999a; 1999b).

The DSD's main function is to organise a private network for the collection and sorting of sales packaging waste, based on plans agreed with each *Land* government and using a mixture of kerbside collection (45% of DSD waste), bring schemes (27%), and combined schemes (28%)[4]. Its operations are financed by a second NEPI, licence agreements between the DSD and manufacturers entitling them to use the DSD's *Grüne Punkt* (Green Dot) logo on their packaging. The principal purpose of this label is to allow consumers to identify packaging belonging to a business participating in the Dual System, thereby helping them to sort waste for collection by the DSD's contractors. In order to obtain this licence, manufacturers pay a fee for each packaging unit bearing the Green Dot, regardless of whether it is recycled, re-filled, or landfilled. The Green Dot therefore also operates as an environmental charge designed to promote more sustainable waste management. The manner in which fees are calculated is complicated but, broadly speaking, takes into account the weight, area and volume of the packaging unit as well as the materials involved (Figure 3.1). Funds from the Green Dot are then used to finance the DSD's operations though, with the exception of plastics, the licence fee only covers collection costs. For more profitable materials, the resale value of reprocessed materials means that subsidies are not necessary; for others, reprocessing costs are allocated directly to the respective packaging producers (Michaelis, 1995). The final link in the recycling chain is provided by the DSD's guarantors. As part of the agreements to set up the Dual System, each material sector established recycling associations to raise funding for recycling and to guarantee the reprocessing of materials collected by the DSD (Michaelis, 1995). A summary of the flow of packaging and funds in the Dual System is shown in Figure 3.2.

The original intention was that the DSD should co-ordinate the actions of all industries involved in the recovery and reprocessing of packaging waste. However, in the early years, this relationship was not particularly co-operative and, as a result, the DSD came close to financial collapse in 1993. Eichstädt *et al.* (1999) attribute this to four problems. First, some manufacturers were free-riding the system either by under-declaring the volume of packaging materials being put through the Dual System or by failing to register and pay for their use of the Green Dot. Second, the licence fees in force at the time contributed little towards the reprocessing of waste packaging. However, public participation in sorting packaging waste rose dramatically in 1992, leading to a shortage of reprocessing capacity in Germany compared with the amount of waste collected. The DSD was unable to find buyers for much of this waste and was forced to export it to other EU states (Waite, 1995; Michaelis, 1995). Not only was this enormously expensive, it also destabilised materials prices throughout Europe by flooding their recycling markets, and undermined the political credibility of the concept of an industry-led recycling system (Lister, 1996; House of Commons, 1997). Third, the high targets imposed by the Ordinance forced the DSD to accept almost any contract it was offered and enabled some disposal firms to charge exorbitant collection fees. Finally, the crisis was compounded when the main guarantor for plastics recycling went into bankruptcy in 1993, citing a lack of reprocessing finance from the DSD as the cause of its insolvency.

To address this problem, a reconsolidation plan was developed by the

Figure 3.1 Green Dot licence fees

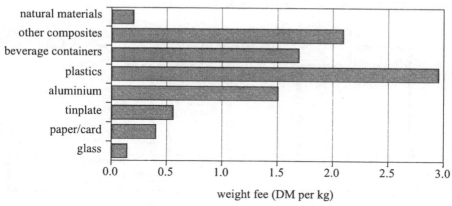

weight fee (DM per kg)

Item Fee in Pfennigs (including statuory VAT)			
Volume item fee		*Area item fee*	
< 50-200 ml and > 3 g	0.1-0.6	< 150-300 cm2 and > 3 g	0.1-0.4
> 200 ml - 3 litres	0.7-0.9	> 300 - 1600 cm2	0.6
Over 3 litres	1.2	over 1600 cm2	0.9

Source: DSD (1998a: 11)

federal Environment Minister (the *Töpfer* Plan). The main elements of this included the conversion of DSD's debts into long-term credit, substantial increases in Green Dot licence fees, and the foundation of a new plastics guarantor *(Deutsche Gesellschaft für Kuntstoff-Recycling mbH (DKR))* by the DSD in conjunction with two German energy suppliers (Eichstädt *et al.*, 1999). As a result of the programme, the DSD has been able to finance more comprehensive collection systems, invest in new recycling technology, and provide increased price support for recycling companies based in Germany. Waste exports have consequently reduced to 19% of packaging waste consumption – yielding large cost savings – and the DSD has stabilised its financial position sufficiently that it was able to announce a 9.5% reduction in the Green Dot fee from January 1999 (*Handelsblatt*, 1997; DSD, 1998a; 1998b). The success of the reconsolidation plan can largely be attributed to three factors; the government's determination not to repeal the Dual System concept, the influence of powerful retailers seeking to avoid the re-imposition of quotas and deposits, and concessions from disposal firms fearful of losing lucrative recycling contracts (Eichstädt *et al.*, 1999). Although the Dual System has subsequently fostered more genuine co-operation between industry sectors,

Figure 3.2 The structure of the German Dual System

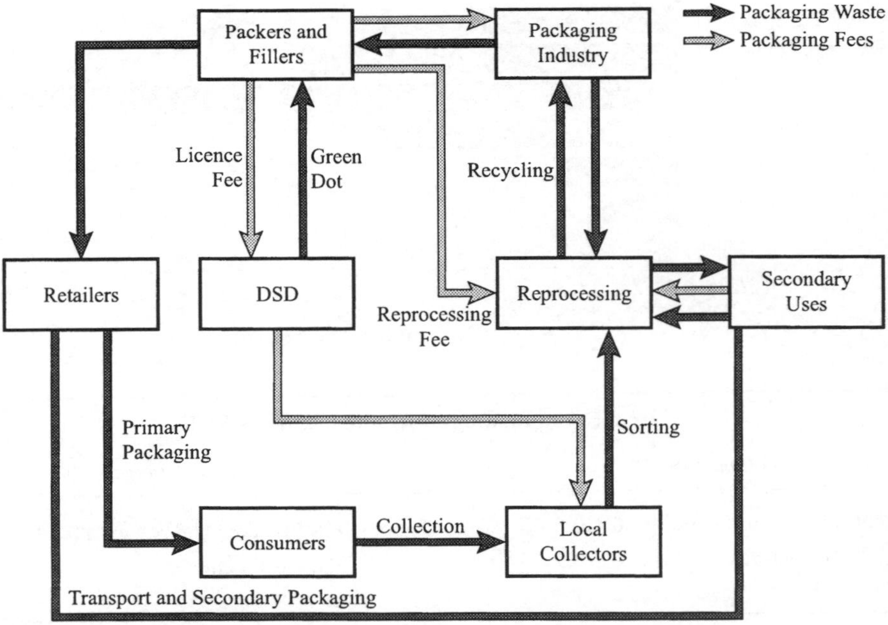

Source: Michaelis (1995: 236)

the main impetus for reform again came from the government's ability and willingness to resort to forceful command-and-control intervention in the German recycling market.

In terms of the changes to the Ordinance brought about by the Packaging Waste Directive, the German government was apparently in an extremely favourable position to adjust to EU requirements. Its legislation and implementing measures were firmly established before the Directive was even agreed, initial implementation problems had mostly been overcome, and the Directive's principal objectives were already being achieved. German industry had also become accustomed to the legalistic and pro-environmental agenda of successive federal governments, which are similar in style to the general tenor of EU environmental legislation (Blüdhorn *et al.*, 1995). The main adaptation pressure caused by the Directive was therefore to eliminate trade restrictions caused by refill quotas and the deferred deposit-refund scheme. However, these issues significantly delayed the German government's efforts to amend the Packaging Ordinance. The *Bundestag's* proposal to dispense with re-use quotas was opposed by the *Bundesrat* on the grounds that they enjoyed public and some industry support in addition to the endorsement of the Fraunhofer Institute (Otto, 1999). As the *Bundesrat* has considerable jurisdiction over the implementation of many areas of environmental policy, it was able to mount an effective campaign against what it saw as undemocratic changes to German policy. Successive compromises were rejected until it was finally agreed that mandatory deposits would only be re-instated if refill quotas were not achieved in two successive years (ENDS, 1998a; 1998b).

This concession could not guarantee the Commission's acceptance of the revised 1998 Packaging Ordinance, however, and subsequent events have inflamed the issue (ENDS, 1997b, 1997c). Following reports by the DSD that refill quotas were not fulfilled in 1997 and 1998, Junior environment minister, Rainer Baake, informed the German industry federation and chamber of commerce that the system would revert to mandatory deposits (ENDS, 2000c). In 2001, the federal cabinet amended the Ordinance to include compulsory returnable deposits of 0.25 Euro on all drinks containers and 0.5 Euro on those exceeding 1.5 litres capacity, effective from January 2002, despite offers from German industry of other environmental concessions in return for relaxed refill quotas (*BMU*, 2001). Whilst these measures have again been defended using life-cycle assessments, the Commission has warned Germany that it may face legal proceedings as a result of this amendment (ENDS, 2000d). The domestic political agenda in Germany has therefore prolonged difficulties in complying with EU environmental law beyond that which might be expected considering the low degree of adaptation required

(Haverland, 2000a). The other major amendment in the 1998 Ordinance to comply with EU recovery targets was a relaxation of restrictions on Energy from Waste (EfW) incineration for transit and sales packaging made from directly 'renewable' materials (Perchards, 1998).

In summary, whilst Germany was instrumental in defining the agenda of EU packaging waste legislation, it has come under intense pressure to harmonise its systems with those of other member states. The dispute raises further questions about the handling of conflicts between the EU's free-trade and environmental objectives, suggesting that the Danish Bottles and the Wallonia Waste Ban rulings may have established some principles but provide only partial guidance on strategies for integrating environmental protection into the wider EU agenda (Bailey, 1999a). The German experience has nevertheless provided important lessons for other member states concerning the implementation of the Packaging Waste Directive. First, it has encouraged governments to be more circumspect about the setting of targets and implementation time frames[5]. Second, it has focused attention on the need to develop viable reprocessing infrastructure from the outset of policy implementation. Third, it has provided insights on the potential trade-offs between environmental effectiveness and economic efficiency in the development of recycling systems (see also Brisson, 1993). As will be shown in later sections, the DSD has proven highly expensive and is seeking to instil greater cost consciousness across all elements of its recycling network (Staudt, 1997; CEC, 2000c). Finally, it is apparent that the German government has maintained a staunchly command-and-control approach towards the implementation of packaging waste policy, despite adopting two policy instruments that are normally associated with more flexible and market-led regulation.

3.4.2 The British Model

Preparations for packaging waste legislation in Britain began soon after the German Ordinance came into effect, with the Department of Trade and Industry (DTI) and Department of the Environment (DoE) commissioning several studies to explore methods for recovering resources from waste (DTI/DETR, 1991; 1992). The main thrust of government policy was to create mechanisms that could achieve the objectives of the Directive 'in a manner which is efficient, equitable and least burdensome' to industry (DoE, 1996a: Ministerial foreword). To meet these aims the DoE sought to reach a voluntary agreement on the format of a business-led scheme for recovering and recycling packaging waste, though it also warned that a unilateral government solution would be introduced if industry failed to produce a suitable plan (Haverland, 1999).

A number of working parties were convened to assist these discussions. Following initial forays by INCPEN and COPAC[6], the government commissioned the Producer Responsibility Group (PRG), a forum of businesses concerned with packaging issues, to prepare a framework plan. The PRG's draft proposals were submitted in February 1994 and established principles which would form the basis of the UK's packaging waste system. The main recommendations of this report were; the introduction of binding legislation which spread the burden of EU targets across all sectors of industry involved in the packaging chain, the creation of a competitive recycling market, and the use of economic instruments to generate incentives for packaging optimisation (Figure 3.3). The sectors covered by these proposals were raw materials manufacturers, converters (manufacturers of packaging), packer-fillers (product manufacturers), and sellers (retailers), with wholesalers taking a combined packer-filler and seller responsibility. On completing its report, the task of fleshing out these proposals was handed to another packaging organisation, VALPAK, and its Working Representative Advisory Group (V-WRAG) (ENDS, 1995a).

Despite the PRG's recommendations, the government initially preferred the administrative simplicity of focusing responsibilities on one sector of the packaging chain (ENDS, 1995b). There was also considerable disagreement amongst industry representatives about the apportionment of sector responsibilities, which caused major problems in agreeing a common proposal.

Figure 3.3 PRG draft proposals, February 1994

i. On current data, recovery of around 58% of the UK's packaging waste is achievable by the year 2000, but not on a voluntary basis. Underpinning legislation is required to assure compliance and to provide the necessary incentive to create business operated schemes to organise recovery and recycling;

ii. All parts of the packaging chain need to be involved, from raw materials manufacturers to retailers, if effective co-operation is to be achieved and recycling costs minimised. It is essential that business sectors co-operate to increase end-use markets for recyclate and to cause investment in new reprosessing capacity while retaining a market led approach;

iii. There is a need for renewed commitment to waste to energy which is more appropriate than recycling for some packaging waste;

iv. There is a need for incentives for minimisation, for example through material-specific charging;

v. There is a need for continuing consumer awareness and participation.

Adapted from: DoE (1996b: pages not numbered)

Even though all companies participating in V-WRAG favoured some form of shared responsibility, retailers supported a single onus on manufacturers and the use of market forces to diffuse costs and responsibilities to other sectors of the chain (ENDS, 1995b), whilst others favoured a variety of multi-point options. The choices considered by V-WRAG and its final proposal are summarised in Figure 3.4.

Figure 3.4 V-Wrag proposals for packaging recovery, 1995

V-WRAG Draft Proposal

 i. **Single Point Obligation:** One point of legal obligation for packaging recovery and recycling (either converters, packer-fillers or wholesalers/retailers), with market forces ensuring all sectors of the chain contribute. This option was favoured for its simplicity and low implementation costs, but raised concerns over the ability of sectors further up the chain to pass on compliance costs to consumers.

 ii. **Omni-Point:** Targeting of 'brand owners' for each product as a means of placing product stewardship obligations with those primarily responsible for generating the packaging. This scheme was seen as complicated and costly to administer.

 iii. **Combined Industry Scheme:** Obligation on the 'first purchasers of packaging for use', primarily packer-fillers.

 iv. **Multi-Point:** Obligation on all sectors of the packaging chain and the allocation of an appropriate share of the recovery targets in relation to the packaging each handles. It emphasised the idea of business joining an industry-wide compliance scheme to manage recovery and recycling activities.

 v. **Equi-point:** A basic requirement on all sectors of the packaging chain to recover packaging waste arising on their premises, with additional obligations upon specific sectors (packer-fillers and retailers to collect household waste, converters in terms of recycled content in packaging, and raw materials manufacturers and reprocessors to reprocess or valorise collected waste).

Source: ENDS (1995b: 37-8)

Shared-Responsibility

 i. Legal duty on all companies to ensure that packaging waste arising on their premises is valorised to agreed levels.

 ii. For packaging supplied further down the packaging chain or to end users, a duty of care to take all reasonable measures to ensure the valorisation of this waste packaging.

 iii. The establishment of a collective scheme (VALPAK) to manage recovery and valorisation responsibilities on behalf of obligated companies.

 iv. Companies choosing to manage their legal duties individually rather than by joining a collective scheme should be required to submit an annual report to the Environment Agency to demonstrate their plans. Failure to do so should be made an offence and the Environment Agency should have the power to require businesses not presenting convincing plans to make good any deficiencies.

Source: ENDS (1995b: 38)

In terms of targets, V-WRAG's preferred a non-quantified duty of care across the packaging chain. This gained widespread industry support but the DoE insisted on clear and binding recovery targets (ENDS, 1995c; 1995d). In the end, the government, mindful of the Commission's notification deadline, forced industry's acceptance of mandatory obligations at a meeting hastily convened by the DoE on 15 December 1995 by threatening to impose a solution (ENDS, 1996). The targets and responsibilities agreed at this meeting are shown in Table 3.3. The DoE, disgruntled with the belligerent attitude of V-WRAG, decided that wider industry consultations and the drafting of legislation should be overseen by a more neutral body and appointed a trusted emissary, Sir Peter Parker, to head the newly formed Advisory Committee on Packaging (ACP). Following its initial work, the UK Regulations were brought before the House of Commons in 1996 (DoE, 1996b).

Though accounts of these negotiations do vary – one anonymous commentator, for instance, claimed that the UK government was 'pathetically anxious not to upset industry and provided no leadership' - the general view is that industry played an important but defensive role in formulating the 'voluntary' agreement. Whilst the *desirability* of a business-led scheme was supported from the outset, industry's wish for a purely voluntary approach provided the British government with insufficient guarantees of compliance with the Directive. Realising that failure to agree a common proposal would risk the complete abandonment of the partnership approach, industry was forced to concede ground. This therefore supports Lévêque (1995) and Whiston and Glachant's (1996) view that many environmental agreements

Table 3.3 Sector targets under shared responsibility

Sector	% of Recovery and Recycling Targets
Raw materials manufacturer	6
Converter	11
Packer-filler	36
Retailer	47
Wholesaler	83

All figures expressed as percentages of the total weight of packaging produced or imported into Britain, with exports deducted. Packaging materials covered by the Regulations are paper/board, glass, steel, aluminium, plastics and, from 2000, wood.

Source: DoE (1996b: pages not numbered)

are not truly voluntary because they are concluded by industry principally to fend off coercive pressure from government. More powerful sectors were nonetheless able to reduce their absolute losses by conceding the legitimacy of the initiative then engaging defensively in the process of policy formulation.

3.4.3 *Producer Responsibility and the PRN System*

In accordance with the agreements reached in 1995, the Producer Responsibility Obligations (Packaging Waste) Regulations 1997 imposed legal responsibility for packaging recovery and recycling on companies deemed to constitute the packaging supply chain (*Packaging Producers*) (DoE, 1997). Thus, firms that only consume rather than provide packaging, such as those in the service and financial sectors, are exempted from direct responsibilities under the Regulations. In order to ensure smooth and cost-efficient progression towards EU requirements, interim targets were established for the period 1998-2001 and the final targets set by the Regulations are towards the minimum permitted by the Directive (Table 3.4). The 1997 Regulations were followed by the Packaging (Essential Requirements) Regulations 1998 (DETR, 1998a), which require packaging to be manufactured in a manner that minimises its volume and weight, and which facilitates its re-use, recycling and recovery. As with the Directive, the 'Essential Requirements' contain no mandatory targets on waste prevention or re-use.

From 1998, all companies with an annual turnover exceeding five million pounds and handling over fifty tonnes of packaging are required to achieve their sector responsibility targets[7]. Using these criteria, the number of obligated UK businesses was initially estimated at 5,000 (Perchards 1998), though the actual number registering with the Environment Agency in 1997 was 3,837

Table 3.4   Targets set by the Producer Responsibility Obligations (Packaging Waste) Regulations, 1997

	Recycling (%)	Recovery (%)
1998	7	38
1999	7	38
2000	11	43
2001	16	52

Source: DoE (1996b: pages not numbered)

(Environment Agency, 1998a). This figure was predicted to increase to over 11,000 in the year 2000 when the turnover threshold was reduced from £5 million to £2 million (DETR, 1998b; 1999a).

Having established legal onus and responsibilities, the next stage was to devise means by which industries could discharge their responsibilities with minimum cost and disruption to normal business activity. To achieve this aim, companies charged with recovering and recycling their packaging have two compliance options. First, they can apply for independent registration with the Environment Agency, Scottish Environment Protection Agency (SEPA), or Northern Ireland Heritage Service, depending on where they are located. Producers choosing this option are required to submit annual waste management plans to the agencies and provide evidence that they have discharged their obligations. Since 1997, approximately 20 per cent of packaging producers have chosen this alternative (Environment Agency 1998a). Second, they can join one of several industry-run *Compliance Schemes*, organisations registered with the Environment Agency to manage producer recycling networks. Producers wishing to take this option must pay membership fees and recovery and recycling charges to the scheme (sometimes termed *materials levies*). In return they secure immunity from prosecution and any practical involvement in the recycling process, as these responsibilities transfer to the scheme. Compliance-scheme charges are generally calculated on the basis of the weight of packaging the producer is required to recover and recycle. Schemes then use the revenue from materials levies to purchase recycling services, making them chiefly a means of aggregating producer obligations into more effective bargaining units. This and administrative simplicity persuaded the majority of the UK's packaging producers that compliance scheme membership was the most cost-effective means of meeting their duties under the Regulations.

The licensing of compliance schemes can be viewed, in many ways, as an extension of the voluntary agreement on sector responsibilities, as it is designed to consolidate and remove the Regulations' more burdensome requirements. However, the arrangement also includes a *de facto* environmental charge on packaging production and use. Thirteen schemes registered with the Agency in 1998 (this number has subsequently risen to seventeen), though VALPAK, the organisation originally charged with developing proposals for industry-led recycling, holds approximately 70 per cent of the market (Environment Agency 1998b). The general principle is that compliance schemes may act on behalf of as many producers as they wish, provided they can demonstrate to the Agency that their packaging recovery services are organised effectively.

The notion of economic instruments is also evident in the physical recycling of packaging waste, where the Environment Agency has licensed *Accredited Reprocessors* to operate what is known as the *Packaging Recovery Note* or PRN scheme. The basic function of PRNs is as a reprocessing certificate. PRNs are used, first, by the Agency to monitor the amount of packaging waste reprocessed in the UK and, second, by producers or compliance schemes as evidence that they have met their recovery and recycling targets. PRN certificates are completed by accredited reprocessors as they physically receive packaging waste, with each certificate specifying the type of material being reprocessed, its weight, and the method of reprocessing used (recycling or incineration with energy recovery) (ENDS, 1998c). The PRN is also an economic instrument because accredited reprocessors are entitled to sell completed PRNs to compliance schemes or producers, which they then use as evidence for their reprocessing returns to the Agency. A synopsis of the packaging waste and funding flows in the PRN system is shown in Figure 3.5.

Several points about the flexibility of this system are important to note immediately. First, provided the reprocessed materials conform to government

Figure 3.5 Funding and packaging flows in the UK PRN System

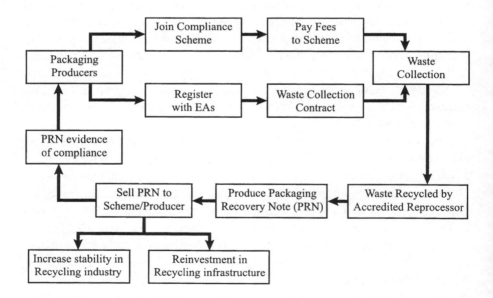

Source: Bailey (1999a: 559) published by Blackwell Publishing

guidelines on items deemed to be packaging (DETR, 1997a; 1997b), PRNs may be produced from either industrial-commercial or household waste. Producers are therefore not necessarily required to recover or recycle their own packaging provided they obtain PRNs for *bone fide* packaging materials. This effectively removes the compulsion for producers to organise their own packaging-waste reclamation, as a considerable proportion of commercial and post-consumer waste is already collected for reprocessing by local authority or other private sector contractors (Turner *et al.*, 1998; DETR, 1998c; 1999b). Furthermore, because the Regulations do not distinguish responsibilities in terms of how packaging waste is procured, industry collection schemes have tended to concentrate on high volume and homogenous industrial-commercial waste, often the by-products of production processes (Bailey, 1999b). This arrangement therefore fulfils industry's request that the Regulations should implement the Directive in a cost-effective manner. Finally, by instigating compliance schemes and PRN charges, the system requires packaging producers to pay for less environmentally damaging methods of waste management but allows them to discharge their duties without any direct involvement in recycling if they so choose.

The PRN system also serves two purposes for accredited reprocessors. First, the additional revenue provides financial support for an industry whose development has historically been hindered by volatile materials prices (Bailey, 1999c; 2000). Second, the expectation is that surplus funds will be invested in new reprocessing capacity, which the government has sought to encourage by allowing PRN prices to be determined by market forces rather than state intervention. According to market theory, PRN prices should respond to the investment requirements of each reprocessing sector, which, in turn, are determined by the incremental increases in recovery targets set by the Regulations. This might suggest that the PRN system is a captive market, as producers and compliance schemes are compelled to buy PRNs or similar forms of evidence in order to fulfil their obligations and, in theory, reprocessors could exploit this situation by inflating the price of PRNs. However, according to government thinking at the time, the balance of market power would be safeguarded, first, by the aggregated bargaining power of compliance schemes acting in the best interests of their members and, second, by competition between compliance schemes and reprocessors (Bailey, 1999b). The rationale of introducing competition to these sectors was that market forces would ensure an efficient allocation of resources and, thus, lower cost implementation of the Directive. By 1999, 210 accredited reprocessors had registered with the environment agencies (by 2001, this had risen to 266), with the most competitive sector being plastics, with 82 market players, followed by paper and glass, with 54 and 20 companies respectively (Environment Agency, 1998c; DETR, 1999a). Monopolistic behaviour is therefore theoretically discouraged by competition between providers of environmental services, creating a system which is market-led, cost-effective and responsive to changing regulatory conditions. However, the system is again underpinned by the understanding that more rigid legislation will be introduced if the PRN scheme fails to achieve EU recycling targets (DETR, 1999a).

3.5   Comparison of the UK and German Models

The policies and mechanisms used by the UK and German governments to implement the Packaging Waste Directive are outwardly very similar in their design and execution. Both countries have framed their policies around making firms more accountable for the environmental stewardship of packaging waste (Fenton and Sinclair, 1996; Sinclair and Fenton, 1997) but used 'voluntary' agreements to grant substantial leeway in the management of these responsibilities. Systems of environmental charges have also been established to apply the polluter pays principle and to fund infrastructure developments, and direct taxation has been avoided in both countries by allowing charges to be determined in part by industry-led organisations. Finally, the diffusion of environmental costs to other polluting parties - generally speaking, the service sector and general public - has been left entirely to market forces (Bickerstaffe and Barrett, 1993).

However, there are fundamental differences in the way the two countries have implemented the Directive and deployed NEPIs. The first key point of divergence has been the importance attached to environmental protection and cost-effective compliance. In terms of environmental standards, the UK government felt that the Directive could impose major costs to industry and took advantage of the legislation's flexibility to introduce the minimum targets possible. Conversely, the German government used the Directive's opt-up clauses to maintain its stringent recycling policies, notwithstanding the stand off over re-fill quotas and deposits. The balance between economic and environmental priorities is also evident in the use of NEPIs by the two governments. Both administrations entered into negotiated agreements with industry to decide the fundamentals of implementation but the UK government's approach was to prioritise the achievement of EU environmental objectives at minimum cost by allowing industry considerable leeway over the apportionment of responsibilities once the principle of binding targets had been established. Those sectors with more established links to government, particularly retailers and manufacturers, were consequently able to gain a comparative advantage by shifting a proportion of their producer obligations towards converters and raw materials manufacturers. However, the German government was far more doctrinaire in terms of the outcomes and the type of system it wished to see emerge from the negotiated agreement and, essentially, the main purpose of the negotiations was to convince industry representatives to accept its position (Haverland, 1999).

In terms of economic instruments, the first distinction between the UK and Germany is the point at which packaging waste charges are levied. Green Dot charges are calculated on the basis of the total packaging produced by DSD members regardless of whether it is recycled, but PRN charges are only payable according to the recovery and recycling rates set by the Regulations. This not only alters the relative cost of the two systems, but also the revenue available for infrastructure development and public education. This distinction is further reinforced by the rates at which recycling charges are set in each country (Figure 3.6). Based on current Green Dot fees (excluding area and volume fees) and average PRN prices between 1998 and 2000, German prices range from between 3.4 times those in Britain (glass) to 14.6 times (plastics) and 25.7 times (aluminium). When both factors are taken into account, it is evident that the Green Dot establishes appreciably stronger financial incentives for companies to reduce the amount of packaging used in their products. At the same time, the UK system contains many unquantified costs, particularly related to waste collection and sorting.

Figure 3.6 Recovery and recycling charges, Germany and UK

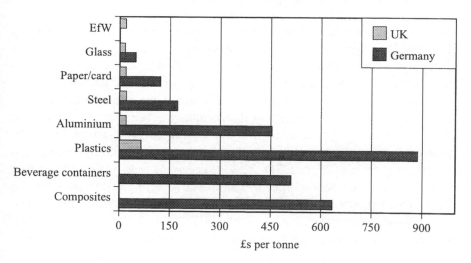

However, the most important distinction is the extent to which market-forces drive the British and German recycling schemes. The PRN system was specifically designed to reap the efficiency benefits of competition and market-based pricing (Beesley and Littlechild, 1983; Gardner, 1996). In Germany, the DSD and its associated guarantors are responsible for co-ordinating all aspects of the recycling system (Michaelis, 1995; DSD, 1998a). Thus, co-operation between industries generally takes precedence over the promotion of competition-led efficiencies. The different priorities in the two countries are further underlined by the methods used to calculate collection and reprocessing costs. Whilst PRN prices are fixed by market forces in response to legislative and competitive pressures, Green Dot charges are centrally determined by the DSD on the basis of life-cycle assessments of packaging's environmental impact and the DSD's operational costs. The German models therefore appears to be more committed to the wholesale integration of environmental concerns into waste management whilst the UK system is more oriented towards compliance with legal obligations at minimum cost. That said, the German government is attempting to instil greater economic efficiency in its recycling networks. The 1998 amendments to the Packaging Ordinance include stipulations that all packaging collection must be reprocessed under competitive conditions and that waste disposal companies must publish their recycling costs annually in order that their competitiveness can be assessed (Flanderka, 1998). This issue is examined further in Chapter four.

3.6   Conclusions

The review of the British and German implementation of the Packaging Directive has revealed wide disparities in the way member states interpret and apply EU environmental law. The German government's implementation model has been broadly framed around the use of command-and-control regulation, with NEPIs deployed purely to overcome implementation difficulties. The first of these was the voluntary agreement to establish the DSD where, in reality, industry's choices were severely curtailed by the government's emphasis on strong environmental targets and determination to steer the organisation of the DSD. This was accompanied by the introduction of the Green Dot financial mechanism at fairly punitive rates to fund the DSD's packaging waste system. Other less formal fiscal instruments have also evolved, such as the creation of the DSD's guarantors to raise additional finance for reprocessing and to provide price support for recycled materials. The general tenor of the German policy has nonetheless remained highly legalistic, prescriptive and regulatory. The British approach to implementing the Directive has been much more experimental and flexible. Whilst similar voluntary agreements and economic instruments have been employed, the British model has stressed the benefits of market-based trading to achieve environmental objectives in a cost-effective manner. Thus, the cost of PRNs is determined principally by supply and demand within a competitive recycling market rather than by government intervention. The producer responsibility agreements were not devoid of coercion but incorporated a sizeable element of industry choice, including the gradual introduction of recycling targets and considerable leeway in the negotiation of sector responsibilities. Throughout all aspects of the implementation of the Packaging Waste Directive, the German government placed primary emphasis on stringent environmental protection, whereas the British government has prioritised the mitigation of economic impacts arising from EU environmental policy. Overall, the UK government has sought a more literal application of the principles of flexibility, cost-effectiveness and choice in the economics literature, whilst German policy has been characterised by an enduring command-and-control mentality and NEPIs have only been introduced to overcome practical difficulties rather than being the centrepiece of its policy.

The key question is whether these differences have made a material difference to the environmental and economic outcomes produced by the Packaging Waste Directive in the UK and Germany. If not, state-led implementation would appear to have achieved higher standards of environmental protection and the harmonisation of legal standards without excessive EU intervention in national affairs. However, if state-led implementation has contributed to significant disparities in national policies, particularly in terms of environmental performance, it is important to understand the implications of this for the coherence of the EU environmental programme. Considering the likely importance of NEPIs to future environmental policies, this raises the question as to whether state-led policy-instrument selection has contributed to an overall convergence or divergence in national environmental policies.

When considering these issues, the first factor to assess is whether the main aims of the Directive have been achieved in both countries. Whilst the ongoing disputes between the Commission and Germany over technical trade barriers have already been reviewed, evaluation is also required of the UK and Germany's progress towards the EU's recovery and recycling targets. Several second-level issues concerning the mechanisms used to implement the Packaging Waste Directive also warrant further examination. First, are the voluntary agreements used to establish the DSD and shared-responsibility systems functioning satisfactorily and equitably? Second, are the Green Dot and PRN mechanisms adequately supporting the expansion in reprocessing and collection infrastructure? Finally, have increased environmental costs produced a discernible effect on industry behaviour, particularly in respect of the Directive's 'Essential Requirements' where mandatory targets have not been set? Whilst the main impetus for increased recovery and recycling in Britain and Germany comes from legislation and economic instruments are used primarily to finance infrastructure developments, both instruments arguably also serve an incentive function promoting the reduction, re-use and recycling of packaging waste. The contribution of the Green Dot and PRN costs to these objectives is therefore also important. In order to explore these questions further, Chapter four examines the development of recycling infrastructure in the UK and Germany, whilst Chapter five discusses the influence of environmental charges on industry's waste management behaviour.

Notes

[1] Articles 100, 100a and 130r are referred to throughout the book although they were amended to Articles 94, 95 and 174 respectively in the Amsterdam Treaty, as the former were in force at the time the Packaging Waste Directive was negotiated.

[2] The prevention of packaging waste is noted as a primary objective in Article 1 of the Directive and as an 'Essential Requirement' in Annex II. Re-use and the development of end markets are dealt with in Article 5 and the preamble respectively. The principle of the waste hierarchy as a general guide to waste policy was adopted in Framework Directive 75/442/EEC and reaffirmed under the amended Framework Directive 91/156/EC (Porter, 1998). The EU's commitment to waste prevention also appears the Commission's *Communication on the Review of the Community Strategy for Waste Management* (COM 96 (399) final) (CEC, 1996a) and the Fifth EAP (Öko-Institut, 1999).

[3] The key exception to this is feedstock recycling, where waste plastics are broken down into their chemical components for use as secondary raw materials in the petrochemical industry, steelworks and refineries. This practice is supported by the DSD since it circumvents some of the difficulties involved in recycling plastics but is criticised by several environmental groups because it is not a 'closed loop' form of recycling (Staudt, 1997).

[4] This information was derived from written correspondence with the DSD. Recycling studies generally agree that public participation is maximised through the use of kerbside collection schemes (Pelletier *et al.*, 1996), except in rural areas, where increased travel discourages public involvement and bring schemes produce less overall environmental impact (Powell *et al.*, 1996).

[5] The schedule to implement the Packaging Ordinance was just 18 months (though Austria opted for a twelve-month implementation programme). By contrast, the UK Regulations were phased in over a four-year period.

[6] INCPEN (Industry Council for Packaging and the Environment) and COPAC (the Consortium of the Packaging Chain).

[7] Sector responsibilities are calculated by multiplying a firm's percentage sector target by the

overall recovery or recycling rate for the year in question. For example, in 1998, raw material manufacturers were required to recover 2.28% of their total packaging consumption (6% sector target x 38% 1998 recovery target).

Chapter 4

Recycling Infrastructure in Britain and Germany

(with Richard O'Doherty and Alan Collins)

4.1   Introduction

The legal, environmental and economic ramifications of the Dual System and UK Packaging Regulations have been the subject of considerable academic debate in recent years[1]. Although it is generally agreed that the German packaging waste system has yet to achieve economic efficiency, opinions are divided as to its environmental merits. Michaelis (1995), Eichstädt *et al.* (1999) and the European Commission (CEC, 2000c) suggest that the Packaging Ordinance has promoted higher levels of recycling in Germany but Staudt (1997) and Flanderka (1998) dispute the environmental worth of the DSD, arguing that recycling was already increasing before the Green Dot was introduced. There is, nevertheless, general agreement that design defects in the UK Packaging Regulations have significantly hampered Britain's efforts to achieve the recovery and recycling targets set by the Packaging Waste Directive. The main foci of concern have been the effects of market-based trading in PRNs and whether such a system can achieve both its economic and environmental objectives (O'Doherty and Bailey, 2000).

It is not the intention to reduplicate the conclusions of these studies but rather to analyse the specific contribution of NEPIs to the success or otherwise of the British and German packaging waste management systems. It was noted in the previous chapter that the two administrations deployed various NEPIs to implement the Packaging Waste Directive. However, although these instruments share common precepts, it is evident that they function in markedly different ways. This chapter considers three functions of NEPIs in the UK and Germany in order to assess their impact on policy outcomes in the two countries. The next section assesses the use of negotiated agreements to establish producer responsibility and industry-led recycling schemes. Following this, the hypothecation of recycling charges to encourage expansion in the reprocessing sector is reviewed. The final section investigates the economic and environmental efficacy of the PRN tradable-permit scheme and offers brief conclusions.

It should be borne in mind throughout the analysis that several factors hamper straightforward comparisons of the British and German systems. Most obviously, the Packaging Ordinance was enacted six years prior to the UK Regulations and, undoubtedly, this has provided German industry with more time to adapt to the operational requirements of the DSD. Cultural factors are also important, particularly Germany's longstanding tradition of government intervention in industry's affairs to promote environmental policy under the system of *Ordnungspolitik* (Egan, 1997). British companies have therefore undergone a steeper learning curve than their German counterparts and allowances must be made for this. That said, with the exception of Small Medium Enterprises (SMEs), which only came under the auspices of the Packaging Regulations in 2000, British firms are no longer in the early stages of this transition. At the time of writing the UK Regulations have been in force for four years and several reviews have taken place to hone legal responsibilities and operating procedures. The NEPIs used to implement the Regulations have also had time to take effect and, as the subsequent discussion will argue, structural difficulties have been revealed which, in some cases, still require government attention.

4.2   Voluntary Agreements in Britain and Germany

Voluntary agreements have become an common feature in environmental policy in recent years, with over 300 agreements in force in the member states, either as unilateral commitments, schemes set up by public authorities, or negotiated agreements between industry and government (Börkey and Lévêque, 2000). In many respects, the suggestion that either the Dual System or producer responsibility are voluntary is slightly disingenuous, as this implies that industry made supererogatory commitments to combat a problem having acknowledged, *a priori*, its moral responsibility to protect the environment. In reality both industry groups were forced to the negotiating table by impending legislation, the need to develop a means of achieving EU targets, and the threat of more constrictive regulation if this could not be achieved through partnership (Nunan, 1999; Haverland, 1999). Both were therefore negotiated agreements to clarify methodologies for implementing EU requirements. Moreover, as Börkey and Lévêque (2000) point out, German constitutional law does not allow the federal government to sign legally binding agreements; consequently, the DSD displays most of the attributes of a voluntary agreement, even allowing for its prescriptive overtones.

The two agreements have now been operating for several years and this itself attests to their robustness as a method for fostering co-operation between industry and government on the implementation of environmental policies. However, a number of problems have also become apparent. The DSD's funding issues have already been identified, as has the German federal government's decision to instate a deposit-refund scheme for beverage containers in response to the DSD's failure to meet re-use targets. However, two further problems have afflicted these agreements; inequities in the apportionment of sector responsibilities (particularly in Britain) and the high number of companies failing to register their recycling obligations with the relevant agencies (free riding).

As noted in Chapter three, the UK system of shared producer responsibility agreed in December 1995 allocated recycling obligations to each sector of the packaging chain according to their responsibility for packaging production and ability to reclaim waste materials. Using this rationale, the largest sector targets were assigned to companies in the latter stages of the packaging chain, namely, packer-fillers, retailers and wholesalers. This part of the agreement was always fragile because of the circumstances in which it was negotiated and began to show signs of stress as the implications of the sector responsibilities became apparent to industry representatives. The first concern was the 83 per cent recovery obligation for wholesalers (whose target was calculated as the combined obligation of packer-fillers and retailers). The intention was that this would capture packaging sold by wholesalers to retailers who fell below the annual turnover thresholds created by the Regulations (DETR, 1998a). However, the wholesaling industry reported to the ACP, the government's advisory body on implementing the Regulations, that the combined obligation was unworkable because wholesalers were unable to assess which of their many small customers fell above and below the threshold. It was therefore argued that the hybrid obligation was onerous and costly to administer and went against the government's declared intention to implement the Directive in the least burdensome manner to industry (ACP, 1998). The government considered this to be a reasonable objection and, in 1998, removed the combined obligation in favour of a straightforward retailer target during its first general review of the Regulations (DETR, 1998b).

Similar discussions took place on the recycling commitment of the converter sector and the turnover threshold levels set by the Regulations. The Packaging Federation, representing the interests of converters, submitted a report in March 1998 requesting a reduction in its recycling quota (ACP, 1998). It argued that the way sector targets had been allocated misrepresented the December 1995 agreement because it overstated converters' responsibility for packaging production and their ability to reclaim waste materials, both of which are largely determined by their customers, typically packer-fillers and retailers. This meant that converters' compliance costs were disproportionately high compared with other sectors of the packaging chain (see Table 4.1). Although the ACP did not entirely support the Packaging Federation's assessment, the DETR agreed to reduce converters' obligation by two per cent and re-adjusted sector responsibilities to those shown in Table 4.1[2].

Table 4.1   Cost assessment and producer responsibilities since 1999[2]

Activity	Turnover £ billions	Costs £ millions	Cost (% of turnover)	Sector Responsibilities (% of packaging)	
				1997	1999
Raw materials	8.2	33	0.4	6	6
Converter	8.5	94	1.1	11	9
Packer-Filler	112.8	237	0.2	36	37
Retailer	123.5	272	0.2	47	48

Source: Bailey (2002: 241) *(Reproduced courtesey of Elsevier Science)*

With regard to thresholds, the British government originally intended that all businesses with an annual turnover exceeding £1 million should be subject to the Regulations from the year 2000. The aim of reducing this threshold from £5 million was to bring more firms into the scheme, thereby increasing the amount of packaging waste available for recovery and recycling, whilst simultaneously allowing smaller businesses more time to prepare for compliance (DoE, 1996a). However, it quickly became apparent that businesses towards the lower end of this scale would struggle with the complexities of the Regulations, particularly in respect of calculating their recycling obligations. As a result, they were also likely to incur disproportionate compliance costs (DETR, 1998b). Although the DETR and Environment Agency had pre-empted such contingencies by developing training material and 'off-the-shelf' software to help companies understand the Regulations, following consultations with the ACP, the DETR agreed to raise the 2000 turnover threshold to £2 million (DETR, 1998b).

The British concept of shared responsibility has therefore not been without its difficulties. Although the agreement has generally held together, individual sectors have sought to reduce disruption to their businesses by arguing that the Regulations imposed unfair costs. In fact, such manoeuvres are common during the early stages of most forms of environmental regulation, as companies seek to engineer a competitive advantage or avert commercial threats. However, global economic pressures and free trade in the Single Market exacerbated these issues in this instance (Lévêque, 1995; 1996a). The converter sector, for example, could not assume that its competitive position was protected by the fact that all British converters were subject to similar compliance burdens, since they are also exposed to competition from other member states and in the wider global economy. Because industry must operate within expanded trade horizons but concurrently observe state laws, international as well as domestic commercial pressures may influence the competitive effects of new environmental policies and need to be considered during policy design (Ekins, 1999). Similar effects have been reported with the phenomenon of carbon leakage linked to the Kyoto Protocol (Hoel, 2001)[3].

In summary, despite numerous disputes over shared responsibility in the UK, there is little evidence that the negotiated agreement has not functioned effectively or that it has been weakened by the various rounds of interest-led bargaining that took place. Considering the scale and experimental nature of the producer responsibility agreement, some retrospective adjustments were always likely and in fact, were provided for by the administration in advance of the policy. Despite sometimes acrimonious disputes, the British government has maintained the agreement using a combination of the legal mandate provided by the Packaging Directive, warnings of more constrictive regulation if industry did not co-operate, and tactical adjustments to the agreements where obvious inequities were identified. Controversies over the apportionment of responsibilities have also dwindled in recent years, reflecting a tactical cycle in industry responses to agreements negotiated in the context of impending legislation. During the initial stages industries have a strong incentive to re-negotiate commitments as adverse competitive effects become manifest. However, as the scope for such manoeuvres diminishes, implementation deadlines approach, or targets increase, the focus switches increasingly away from tactical manoeuvring towards ensuring compliance with the practical requirements of the legislation.

Free riding has been a persistent problem for both the UK and German negotiated agreements, however. One of the main reasons for the Dual System's financial difficulties in 1993 was the large number of companies using the Green Dot without paying licence fees to the DSD (Eichstädt *et al.*, 1999). The free-rider factor was also instrumental in the DSD's failure to meet its re-use targets, which prompted the re-imposition of deposit-refund quotas by the federal government. Although the German administration acted in accordance with its view that industry had collective responsibility for meeting DSD targets, the DSD has appealed to local authorities, who are permitted under *Länder law* to impose fines on free riders, to take a tougher line in their area (Perchards, 1998). Few authorities have acted, however, and those that have tended to impose minimal fines. Another way the DSD can combat free riders is to take out private litigation against companies using the Green Dot, a licensed trademark, without permission. The DSD has acted against over 750 companies, though the legal process has often proved to be expensive, time consuming and not always effective (Perchards, 1998).

In order to bolster the sanctions against free riders, the federal government included new enforcement powers in the 1998 amendment to the Packaging Ordinance. The original Ordinance did not require companies that chose to remain independent of the DSD to submit details of the methods being used to fulfil their obligations. However, the revised legislation requires non-DSD members to provide the federal environmental protection agency (*Umweltbundesamt*) with documentary evidence of the amount of packaging being placed on the market and recycled (*Umweltbundesamt*, 1998). Industry has also taken a role in self-policing the agreement. Some retailing chains, for example, now deduct Green Dot fees from payments to suppliers who cannot prove that their DSD returns have been independently audited. The deductions, which are based on the sales value of affected goods rather than the type and weight of packaging involved, range from 0.5 per cent of sales price for textile products to 2.5 per cent for food and detergents (Perchards, 1998). There are indications that free riding is diminishing as a result of these interventions; 1800 new companies were recruited to the DSD in 1996 and a further 1568 joined in the year 2000 (DSD, 2001a). However, the DSD continues to suffer from free riders, with the inevitable side effect of making the system more expensive for existing members. Estimates of the annual cost of free riding ranged from Deutsch Marks (DM) 600-800 million (£188-251 million) (Perchards, 1998).

Whilst many factors have contributed to the German free-rider problem, the federal government's initial reluctance to deviate from the principle of collective responsibility and make the prosecution of offending companies a first line of enforcement seem to be particularly important. Börkey and Lévêque (2000) remark that where individual liability is established in a negotiated agreement, as occurs with most Dutch covenants, the free-rider problem is all but eliminated. However, the British experience indicates that even this may not be sufficient if there is inadequate enforcement of the agreement (Bailey, 1999c). Whilst the introduction of shared responsibility and compliance schemes theoretically lessened the temptation to free ride the Regulations by reducing the administrative and financial burden on individual sectors, approximately 34 per cent of British companies nonetheless initially decided to register independently. This number has subsequently fallen to approximately 20 per cent (ACP, 2001). The DETR decided that these companies should submit annual compliance plans using a standard reporting format similar to that employed by compliance schemes (DETR, 1998b).

This still left two other avenues for free riding; total evasion (failure to register with a compliance scheme or the Environment Agency), and under-declaration of the amount of packaging waste being produced by the company. The calculation of individual obligations proved particularly contentious because of uncertainties over what materials should be classified as packaging. There were lengthy and arcane debates over whether such items as coat hangers and plant pots constituted packaging, with industry claiming that their inclusion would add phenomenally to compliance costs (Daily Telegraph, 1997). In order to settle such disputes, *Ready Reckoners* and *User's Guides* were introduced to clarify responsibilities and definitions of packaging (DETR, 1997a, 1997b). The ACP (1998) also reported that many companies had significantly under- or over-declared their early returns because they had simply misunderstood the guidelines.

However, detecting and successfully prosecuting legislation dodgers has proven difficult. Although producers pay an annual registration fee to the environment agencies, either directly or through compliance schemes, to fund their monitoring efforts, it was acknowledged from the outset that not every company could be inspected each year. In fact, monitoring of the Regulations has become heavily paper-based, relying on analysis of producer and compliance scheme returns, and annual visits to a sample of firms. The Environment Agency's handling of enforcement has been roundly criticised, with reports of hundreds of companies not registering, perfunctory 'drive-by' inspections, and a general lack of resources devoted to monitoring activities (ENDS, 1998d, 1998e). The government has also admitted that the legislation contains several major ambiguities, particularly in how legal duties are defined, which have militated against successful prosecutions (ENDS, 1998f, 2000e). In one of the most blatant loopholes, the Regulations provide that when a company's application for registration is granted, it shall be deemed to have registered from the beginning of the relevant year. This period of grace has allowed unscrupulous producers deliberately to register late in the annual reporting cycle and still remain immune from prosecution. Despite some convictions, the preferred avenues for increasing registrations have been the issuing of warning letters, education programmes for the SME sector and initiatives to help companies calculate their recovery commitments[4].

Overall, the evidence from the Packaging Waste Directive suggests that governments are still learning about the design of negotiated agreements as an instrument of environmental policy. Although there is general acceptance that closer co-operation between industry and government is indispensable, putting the technique into practice requires an ability to reconcile a multitude of conflicting interests whilst retaining a clear focus on desired policy outcomes. Consequently, negotiated agreements may undergo periods of mutability and uncertainty during their early stages. The UK and German experiments with producer responsibility nonetheless highlight useful lessons for negotiated agreements in other areas of environmental policy. First, the success of negotiated agreements is often determined by the amount of regulatory threat the government has at its disposal. Without an adequate stick or a well-established corporatist approach to environmental policy, as exists in the Netherlands, industry groups may use their technical expertise or economic arguments to resist or distort the government's environmental aims (Wilson, 1996; Haverland, 2000a). In the case of the Packaging Waste Directive, EU and national legislation severely constrained industry's negotiating options beyond securing flexibility in the way regulations were interpreted. Second, maintaining flexibility of policy options can be pivotal in ensuring government objectives are achieved. For example, both the UK and German governments reserved the right to introduce new command-and-control measures if industry failed to accede to their wishes; this enabled the German government to pursue its policy of encouraging packaging re-use despite considerable industry antagonism. The British government has meanwhile exercised its discretion to remove obvious inequities from the shared responsibility agreement, thus helping to preserve a consensus in favour of the accord despite industry's unease about the Regulations.

Finally, enforcement of the negotiated agreements has been the biggest challenge for both governments. Although voluntary agreements which incorporate individual liability should theoretically reduce monitoring costs compared with command-and-control legislation assuming the majority of businesses are prepared to fulfil their obligations, enforcement responsibilities and resources still need to be clearly defined to restrict free riding (Börkey and Lévêque, 2000). An ambiguous legal framework has hindered the Environment Agency's enforcement efforts, lax reporting requirements have caused the DSD major difficulties and, in both countries, the sheer scale of monitoring thousands of businesses has stretched the capabilities of enforcement agencies. These problems are not unique to voluntary agreements but are nonetheless important considerations in the design of policies that use this technique.

4.3   Hypothecation and The Expansion of Recycling Infrastructure

It is sometimes tempting to overlook the fact that the implementation of environmental policies is a complex process and that setting targets does not inevitably lead to the achievement of desired policy outcomes. Equally, effective policy implementation frequently requires changing attitudes and behaviour that, more often than not, have some negative consequences on the immediate interests of particular sectors of society. Recycling is no exception to this rule and requires immense co-ordination of activities and actors (Sinclair and Fenton, 1997). Four main tasks can nonetheless be identified. First, governments must ensure that there is sufficient reprocessing capacity in the national system. Here it is important to bear in mind that although the Packaging Waste Directive sets targets for both the recovery and recycling of packaging waste, it also allows states to exceed their recycling quota at the expense of incineration provided the overall recovery target is met. This option was favoured by the German government, which viewed incineration as failing to contribute towards a 'closed loop' of materials consumption and re-use. By contrast, the British administration saw energy from waste (EfW) as an important part of its waste strategy, not least because of the relative ease and lower cost of incineration. The second task is to develop commensurate collection infrastructure, since materials must obviously be collected before they can be reprocessed. Third, increased public participation in collecting and segregating household waste is needed. Although the UK and German recycling schemes are both ostensibly industry-led, a significant proportion of packaging waste ultimately comes from household waste and systems for its retrieval must be developed. Finally, if recycling is to remain viable, a healthy market for reprocessed materials is needed. The German and UK governments sought to find policy mechanisms that addressed each of these tasks and which fostered co-operation between the sectors of the recycling chain (broadly speaking, packaging producers, reprocessors, waste collection authorities, the general public, and purchasers of recyclate).

A combination of primary and secondary data sources were used to conduct the assessment of these activities. The secondary data were derived from government documents, specialist press releases, consultants' reports and academic commentaries. Primary data were obtained from a survey of UK accredited reprocessors. A corresponding survey of the German reprocessing sector was not necessary because relevant information was already available from secondary sources. 133 UK reprocessors were contacted for the survey, achieving a response rate of 36.1 per cent. Qualitative comments from the survey were used to highlight producer, compliance scheme and reprocessors' opinions on the collection and reprocessing of packaging waste as well as the merits of the PRN system. The DETR conducted a similar review in 1999 which provides comprehensive quantitative data on the UK reprocessing sector (DETR, 1999a).

4.3.1 Reprocessing Capacity

The primary function of the Green Dot and PRN mechanisms is to raise finance from packaging producers for investment in recycling infrastructure in accordance with the polluter pays principle (Eichstädt *et al.*, 1999; Bailey 1999b). In Germany hypothecation is co-ordinated by the DSD, whereas in Britain investment is generated through market exchanges between packaging producers, compliance schemes and accredited reprocessors, with the Environment Agency serving a supervisory role. From an initial inspection it would seem that both schemes have stimulated significant increases in reprocessing capacity. Of the DSD's sales income of DM 4.0 billion in 2000 (£1.26 billion), DM 3.50 billion (£1.11 billion) was spent on waste management and recycling services (DSD, 2001a)[5] and in 1998, an estimated £56 million of PRN revenue was made available to UK reprocessors, though the amount actually spent on recycling networks remains a matter of conjecture (DETR, 1999a; ACP, 1998)[6].

In Germany, this revenue produced a six-fold increase in recycling between 1992 and 1998, with a recycling rate for sales packaging of 84 per cent in 1998 (Table 4.2). Waste exports have also dropped dramatically over this

Table 4.2 Recycling of packaging in Germany

| | Thousands tonnes recycled | | | | Sector Responsibilities (% of packaging)[a] | |
	1992	1995	1998	2000	1998	2000
Paper	306	1255	1416	1506	89	167
Glass	542	2572	2705	2664	87	91
Aluminium	1	32	43	41	83	95
Steel	29	259	345	318	73	112
Plastics	41	504	600	570	69	93
Composites	5	297	345	375	-	67
Total	924	4919	5454	5474	84	103

[a] Commission data allocate composites between the main materials according to their market share in 1995 (55% paper, 39% tinplate steel, 3% plastics, 3% aluminium).

Sources: DSD (1998a, 1999a; 1999b; 2001a): CEC (2000b, p. 135).

period to below 19 per cent. When transport and secondary packaging are included, the 1998 recycling rate was lower at 64 per cent, though it remains towards the maximum required by the Packaging Waste Directive (Perchards, 1998). It is also noteworthy that the sales recycling figures for paper and steel tinplate for 2000, shown in Table 4.2, are slightly misleading, as they appear to exceed 100 per cent. This is because the Mass Flow Verification produced by the DSD's guarantors calculates packaging production on the basis of collection returns from waste management companies and includes waste belonging to unlicensed companies. The data nonetheless shows that the Green Dot has supported significant growth in domestic reprocessing capacity notwithstanding the DSD's early problems with waste exports.

The DETR survey reveals a more varied picture for reprocessing in the UK (DETR, 1999a). Table 4.3 shows predictions of reprocessing capacity for each materials sector and highlights an average increase in capacity of 22.0 per cent between 1998 and 2001, and 27.4 per cent when EfW is taken into account. These figures reduce to 13.3 per cent and 16.8 per cent when wood packaging is excluded, as actual reprocessing data for wood are only available for 2001 and, therefore, are not included in the base figures for 1998. The DETR then compared the estimated reprocessing capacity required in each sector to achieve the Directive's 50 per cent recovery and 25 per cent recycling rates, as well as that required to meet the Directive's minimum requirement of 15 per cent recycling per material. The information for each

Table 4.3   1999 predictions of UK recovery and recycling in 2001 (thousands tonnes)

	Estimated packaging production 2001	Actual reprocessing capacity 1998	Estimated reprocessing capacity 2001	Capacity required 50% recovery	Capacity required 15% recovery	Capacity required 25% recovery	Balance against 25% recovery
Recycling							
Paper	4308	1888	1921	2154	646	1077	844
Glass	2200	658	730	1100	330	550	180
Aluminium	111	15	53	56	17	28	25
Steel	735	183	235	368	110	184	51
Plastics	1912	126	212	956	287	478	-266
Wood	1300	n/a	350	650	195	325	25
Total	10566	2870	3501	5284	1585	2642	859
EfW		448	726				
Total recovery		3318	4227	5284			
Shortfall		n/a	1057				

Source: DETR (1999a: 21) and DETR (1999c)

packaging material was calculated for recycling only, whilst the incineration figures are amalgamated for all materials. The DETR concluded that every sector with the exception of plastics is capable of exceeding 25 per cent recycling, reducing the need for EfW to achieve 50 per cent recovery. Even allowing for this, in 1999 the recovery of packaging waste in Britain in 2001 was predicted to fall 1.06 million tonnes below EU requirements because of deficits in incineration and plastics recycling.

The DETR also provided a qualitative assessment of the position in each material sector, highlighting factors that have inhibited their expansion (DETR, 1999a). It noted, for example, that paper and glass recycling are strongly affected by international commodity prices and are in competition with cheap and plentiful virgin materials (see also Hanley and Slark, 1994). Glass recycling is also hampered by the variable quality of collected materials, caused by consumers mixing different colours of glass at bottle banks. Finally, the DETR acknowledged a severe mismatch between supply and demand for green glass in the UK. Although in recent years British consumers have become increasingly fond of red wine – which is traditionally sold in green bottles – the economy produces relatively few products that utilise green glass. As a result, there is only modest demand for green glass recyclate and the majority is currently shipped to Latin America for secondary uses.

The DETR review provided more extensive commentaries on the shortfalls in the EfW and plastics sectors. The performance of EfW was considered especially disappointing bearing in mind its importance to government waste plans. In *Less Waste More Value* and *A Way with Waste*, two papers circulated in 1998 and 1999 to consult on strategies for sustainable waste management in England and Wales, the recovery of energy from waste to conserve other fuel sources was seen as particularly meritorious (DETR, 1998c; 1999b). The DETR nonetheless noted that ten new incineration plants had been accredited, three were under construction and a further eight were awaiting planning permission (DETR, 1999a). It was anticipated that these would contribute an additional 120,000 tonnes of reprocessing capacity for packaging waste each year[7]. The main obstacle for EfW, in the DETR's opinion, was public resistance to the location of incinerators near their communities (NIMBYism - Not in My Backyard) (DETR, 1998b; see also Goldman, 1996; Elliot, 1998). It added that if the under-capacity in EfW was not rectified or cross-subsidised by over-capacity in recycling, compliance schemes and producers might be forced to export packaging waste for reprocessing. In fact, for a variety of reasons, a niche market in Packaging Export Recovery Notes (PERNs) has already emerged and is expected to expand further, particularly for plastics (Materials Recycling Week (MRW), 1998a; 1999a). Whilst this might circumvent the immediate problem of meeting EU targets, it contravenes the proximity principle established by the Wallonian waste ban and puts UK policy at odds with the spirit of EU environmental law.

The recycling of plastics is naturally disadvantaged by the fact that most plastics are composite materials, which are difficult and costly to reprocess (Bailey, 1999b; 1999d). This, along with the poor economics of plastics collection (for example, over 33,000 half-litre bottles are required to collect one tonne of material) appear to outweigh the fact that plastics is a highly competitive reprocessing sector, which should theoretically make it more cost-efficient. However, the poor prospects for plastics recycling have dissuaded major manufacturers of plastics from registering as accredited reprocessors[8]. Producers who are able to fulfil their obligations from other materials have also tended to avoid plastics recycling, though the Directive's 15 per cent recycling target for all packaging materials from 2001 has increasingly restricted this option (MRW, 1999b; 1999c).

One of the more general lessons to be gained from this experience is that uniform market-led solutions may not be appropriate for all sectors in relation to particular environmental problems. Although markets have the ability to react speedily and efficiently to price signals and regulatory pressures, if these do not make the provision of an environmental service profitable in the medium to long term then more players will leave the market than join. This has certainly been the case for the reprocessing of packaging waste, where market players have no direct legal obligations under the Regulations. As the technical and economic issues vary markedly between different sectors, it would seem sensible to tailor solutions to the characteristics of each market, using sector-specific regulation to ensure that implementation stays on course. This latter approach has been followed more closely in Germany, where a supplementary fee is levied to contribute towards reprocessing costs and to support the price of recycled plastics. Consequently, plastics recycling in Germany has exceeded the minimum requirements of the Packaging Directive. There are also indications that the Commission may propose sector-specific targets in its amendments to the Packaging Waste Directive covering the period to 2006 (ACP, 2001).

4.3.2 Collection

The principal function of the German Dual System is to organise the collection of post-consumer packaging waste in Germany and, as has already been indicated, the majority of its budget is devoted to the organisation of sub-contracted kerbside collection schemes. Academic studies have consistently shown kerbside collection to be the most effective way of maximising public participation in recycling. At the same time, such schemes tend to be more expensive per unit of recycling than bring schemes such as bottle banks, especially in rural areas where transport costs are higher (Pelletier *et al.*, 1996; Powell *et al.*, 1996; Barrett and Lawlor, 1997). Nonetheless, German consumers collected over 5.7 million tonnes of waste in the year 2000 – 78.3 kilograms per person – an increase of 0.8 per cent on the 1999 figure (DSD, 2001b). Whilst many of the DSD's collection contracts were initially set at disadvantageous rates because of pressures to meet the requirements of the Packaging Ordinance, there is evidence that market power is now more evenly distributed. As noted in Chapter three, the 1998 revision to the Packaging Ordinance requires all waste collection and disposal companies to publish their recycling costs annually in order that their competitiveness can be assessed (Flanderka, 1998). Moreover, although the Dual System remains the most expensive recycling system in Europe both in absolute and relative terms, high cost does not mean that the Green Dot has failed in its environmental function. The system's economic inefficiencies have been caused by the monopolistic nature of the DSD and the absence of market pressures from the German reprocessing system but not by the introduction of a hypothecated environmental charge *per se*[9].

In Britain the DETR condemned what it saw as unacceptably low private-sector investment in collection facilities and urged reprocessors to make further efforts in this area (DETR, 1999a). Whilst the ACP believes that the majority of the 2001 target can be met through the recovery of commercial-industrial waste, the Regulations have themselves stymied a significant source of waste collection by not providing local authorities with automatic access to PRN funds (ENDS, 1996; ACP, 2001). Lack of readily available finance has restricted councils' ability to develop kerbside or other similarly user-friendly waste collection schemes despite the fact that reprocessors obtain a considerable portion of waste materials from local authority contractors[10]. Overall, co-operation between reprocessors and local authorities has been very uneven, with some reprocessors maintaining links with over 300 councils and others virtually none (Bailey, 2000). Again some sectors have been more adversely affected than others. Generalising slightly, most local authorities have well-established schemes for collecting glass, aluminium, paper and steel but the number of schemes is significantly lower for plastics, again because of the poor economics of plastics recycling and lack of demand for the material.

Notwithstanding this, the British government's indignation at the lack of expansion in waste collection seems slightly bizarre, given that it sanctioned a hypothecation scheme which prevented a major link in the recycling chain from gaining direct access to PRN revenue (ENDS, 1996). Even without the benefit of hindsight, the administration's faith in the ability of market forces to promote inter-sectoral co-operation was optimistic and such problems should have been foreseen. In order to correct this oversight, the DETR instructed reprocessors to submit annual reports detailing the proportion of PRN revenue being spent on collaboration with local authorities. It also briefly toyed with the idea of separate recycling targets for household and industrial waste to stimulate post-consumer waste collection (DETR, 1998b). This was dropped, however, because it was felt that it increased the complexity of the Regulations at a time when many firms were still to adjusting to their basic responsibilities (Bailey, 2000). The latest review from the ACP has also suggested that greater co-operation between compliance schemes, reprocessors and local authorities will be essential to achieving the EU's recycling targets for 2006 – which are thought to be approximately 60-65 per cent – as the spare capacity in commercial-industrial waste streams is dwindling (ACP, 2001).

In terms of the lessons from this experience, whilst the PRN system has performed poorly in relation to waste collection, again its deficiencies do not stem from any fundamental defect in the concept of hypothecation. However, the Conservative administration applied the logic of the market in a somewhat dogmatic fashion when designing the PRN scheme, when the immediate priority should have been to promote inter-sectoral co-operation. The situation therefore demanded a more pragmatic approach and more careful balancing of statutory environmental responsibilities against desired economic outcomes.

4.3.3 Public Participation

Reviews of the German Packaging Ordinance have generally portrayed the federal government as legalistic and authoritarian in its dealings with the packaging industry (Bailey, 1999a; Haverland, 1999). However, garnering public support for, and participation in, recycling required an altogether more suasive approach. Many aspects of the DSD, such as the Green Dot and kerbside collection schemes, were already designed with awareness raising and ease of use in mind and have been complemented by high-profile annual recycling-awareness days. During its 1997 event, over 300 recycling facilities opened to the public, attracting 830,000 visitors (MRW, 1998b). As a result of these activities and reams of other promotional literature, the DSD reports that 61 per cent of German citizens support the idea of the Green Dot (an increase from 39 per cent in 1993), whilst 95 per cent sort their packaging waste for recycling (DSD, 1998a; 1998b).

The British model is less obviously geared to towards public involvement even though the 1998 review of the Regulations stressed the importance of public education for the long-term success of the recycling system (DETR, 1998b). Legal responsibilities remain heavily focused towards producer responsibility and, although there is a link between PRN revenue and public participation through collaboration between reprocessors and local authorities, it remains convoluted and has not flourished. Some compliance schemes have developed consumer awareness programmes; VALPAK, for example, uses its contracts to provide funding for public education, whilst Biffpack, Properpak and Recycle UK have invested in kerbside recycling as part of their function as waste management companies (Bailey, 1999b). However, the DETR review conceded that: 'it is not clear that these are necessarily policies with real impact' (DETR 1998b: 41).

The DETR responded to what they perceived as compliance schemes' apathy towards public involvement by demanding that they devote more resources to education and submit annual reports detailing their expenditure in this area. The government also briefly considered more coercive measures. In *Less Waste More Value*, it hinted at the introduction of direct weight-based fees for household waste collection to replace the current flat-rate charges incorporated into council taxes (DETR, 1998c). The intention was to establish a direct link in the public mind between waste production and the environmental cost of its management. This plan was subsequently dropped because it was felt that direct charging would have a disproportionate impact on less affluent households (ACP, 1998)[11]. The government also accepted that individual households had limited influence over how products were packaged. The steps taken by the British government to promote public participation have therefore consisted of minor adjustments to industry responsibilities and radical alterations to the Regulations have been deferred on the grounds that further monitoring is required before informed decisions can be made on how best to reform the system.

The experiences of the Green Dot and PRN schemes demonstrate some interesting general points about hypothecated taxes and charges. The traditional objection to hypothecation is that it introduces excessive rigidity into public income and expenditure (Spackman, 1997). However, this has largely been obviated in both countries by the creation of a closed-loop funding mechanism between sectors involved in the production and recycling of packaging waste. The second objection to hypothecation is that, in many cases, there is a dislocation between the amount of revenue raised and that needed to meet particular environmental objectives. If these do not naturally coincide, Smith (1997) suggests, economic inefficiencies may result or insufficient resources may be made available for environmental protection.

In Germany economic inefficiencies have clearly occurred because Green Dot charges were set on the basis of rather myopic environmental criteria with little obvious appreciation of economic exigencies. Calculations based on various official data suggest that each tonne of materials recovered in Germany has cost £228, compared with a preliminary estimate for the UK of £150 per tonne (DoE, 1996a; DETR, 1999a; DSD, 2001a). Taylor Nelson Sofres indicate that the differential may be greater still, with a cost of £196 for each tonne of household waste recovered in Germany and £17-24 in Britain (CEC, 2000c). ACP data even imply a reprocessing cost of £8.58 per tonne in the UK (ACP, 2001). Whilst other factors need to be taken into account, such as the exclusion of non-sales packaging from the DSD, lack of data on UK waste collection and sorting costs, and the different emphases on EfW and recycling in the two countries, there would appear to be a basic trade off between economic and environmental effectiveness in the two hypothecation schemes.

In summary, although the British and German hypothecation schemes have both had their detractors, there is little evidence to suggest that the technique cannot be used to produce environmentally and economic efficient solutions to environmental problems (Bailey, 1999c). The German approach has proven expensive purely because the government has remained predisposed towards achieving high rates of recycling and re-use and shown less urgency in addressing the scheme's economic inefficiencies. By contrast, the British hypothecation scheme placed great credence in the idea that profit-driven trading would inevitably lead to an efficient allocation of resources according to environmental criteria. Neither government appears as yet to have struck a balance between these prerogatives, though both have introduced incremental command-and-control measures with this intention. The UK government has nonetheless remained eager to retain the more desirable features of the market-led system but whether the current reforms to hypothecation arrangements will allow it to replicate the successful elements of the German model whilst avoiding the latter's costs remains an open question. In terms of overall conclusions, therefore, hypothecation has proved to be a useful ancillary mechanism for overcoming practical and financial difficulties arising during the implementation of environmental policies. At the same time, care is needed to ensure that the regulatory regime maintains a reasonable balance between environmental and economic effectiveness. It is also worth remembering that *post hoc* adjustments to existing systems usually incur higher transaction costs than changes during initial policy design.

4.4   PRNs and Tradable Permits in Packaging Waste

The most novel aspect of the British recycling system and its point of greatest divergence from the German model was the decision to introduce tradable PRNs for packaging waste. To re-cap briefly, accredited reprocessors are required to complete PRN certificates for each tonne of packaging waste recovered or recycled. These are then sold to producers or compliance schemes to provide evidence of having discharged their recovery and recycling obligations. The proceeds from PRNs sales are then intended for re-investment in collection and reprocessing infrastructure, public awareness schemes, and the development of markets for recyclate. PRN prices are determined by supply and demand pressures between the two sectors, which are in turn underpinned by UK and EU legislation (Bailey, 1999b; O'Doherty and Bailey, 2000).

This arrangement was initially hailed by the UK government as minimising the need for direct intervention in reprocessing markets. Many producers and compliance schemes were nevertheless sceptical about the PRN scheme and became openly hostile as its structural weaknesses became apparent. The main brunt of criticism was the government's theoretically elegant but untried market-led approach to the development of recycling infrastructure. Throughout the discussion, it should be remembered that accredited reprocessors are not legally responsible for meeting the government's recovery and recycling targets. Neither are they obliged to register for PRN accreditation though those declining this opportunity forfeit the right to PRN income (Bailey, 2000). The involvement of reprocessors in the PRN scheme is therefore principally market led and profit motivated.

The first accusation of malpractice came from compliance schemes and producers. Though the PRN market promotes competition between reprocessors, the Regulations established an annual deadline at the end of each January when producers and compliance schemes must provide the environment agencies with PRN returns. Producers argued that this created an artificial boundary in trading and a natural surge in PRN demand toward the latter part of the annual cycle, which some reprocessors were exploiting by restricting the supply of PRNs during the early part of the cycle in order to push up their price (ENDS, 1998c). This, they claimed, increased producers' compliance costs for reasons unrelated to the objectives of the Regulations. One producer association interviewed hinted at the possible causes of this difficulty:

There are concerns that this market mechanism could be abused by individuals or organisations seeking to profit by trading in PRNs, e.g. by stockpiling until the compliance date is imminent. This would merely increase the cost to obligated companies and make no contribution to the growth of recycling. This concern would not arise if the mechanism to demonstrate compliance were to be separated from the mechanism needed to inject funds into the system as is the case with most EU member states (producer association).

Some compliance schemes have added that the very notion of charging for PRNs is counter-productive for recycling. Under the PRN system, producers that collect and deliver their own packaging waste for reprocessing are obliged to pay the same price for PRNs as those that discharge their responsibilities entirely through PRN purchases without making any attempt to increase the physical recovery of waste materials (O'Doherty and Bailey, 2000). This, it was argued, created a financial disincentive for businesses to set up their own collection schemes:

> An effective PRN system should have provided free proof of recycling to the originators of the waste. Those segregating waste and making it available for reprocessing should then have benefited from the value of any Tradable Permit. This notion was ignored and instead a system whereby reprocessors charged fees to reprocess material has sprung up in its place (compliance scheme).

These matters were referred to the DETR by the ACP, whose main role became to consult with industry on the application of the Regulations (ACP, 1998). It noted that the situation was being inflamed because the system contained no restrictions on who could obtain PRNs. This led to rumours that parties with no connection to the recycling industry were acquiring PRNs specifically to speculate on their value, again using the January reporting deadline as leverage. The producers' underlying grievance was that the Regulations forced them to trade with reprocessors regardless of their abatement costs, rather than allowing the choice of whether or not to enter the market, as would occur with most permit schemes. Sir Peter Parker, chair of the ACP, concurred that the dislocation between the power held by reprocessors and their level of legal responsibility was causing problems and suggested that inadequate regulation had created inconsistencies in the market system (ACP, 1998; Bailey, 2000).

The immediate question is whether the creation of conditions where profiteering is theoretically possible is in fact symptomatic of a dysfunctional tradable-permit system. It might be argued, for instance, that reprocessors who withhold or sell PRNs to speculators are merely utilising market factors to maximise the revenue available for recycling investment. Such actions may even force market prices upwards and generate investment funds more quickly. Under this scenario, high permit prices should prevail in the early stages of the scheme followed by a fall in value as the reprocessing market becomes saturated (Cummings *et al.*, 2001). Whilst this might initially discourage the segregation of waste by producers – which is only likely to accelerate when legislative targets are raised – it is not conclusive evidence of a failing system provided materials are made available from other sources, such as domestic waste. It should nonetheless be remembered that the PRN system also sought to provide producers with cost-effective means of implementing the Regulations, especially during their early stages. But if reprocessors and speculators are able to exploit a captive market when PRNs are in short supply because of insufficient materials, monopolistic pricing and economic inefficiencies may then occur. Recognising these weaknesses, a number of options have been suggested to redress the situation, all of which seek to spread the demand for PRNs more evenly. This could be achieved, for example, by using four cut-off points throughout the year, with producers being allocated reporting dates which smoothed demand for PRNs (O'Doherty and Bailey, 2000). More ambitiously, monitoring of compliance could take place over an extended time period using 'banking and borrowing' from year to year, a system employed in other permit schemes[12]. Given the concerns at the time over whether the UK would meet EU recycling targets, neither option has yet been reviewed in detail. Recent research has nonetheless suggested that a number of interested parties have sought to influence the government's future decisions on cut-off dates (O'Doherty and Bailey, 2000).

In fact, the evidence to confirm widespread profiteering has not been conclusive. If mass speculating had occurred, PRN prices would be expected to rise markedly towards each January deadline, as hoarding caused a bottleneck in PRN demand. Figures 4.1 and 4.2, showing PRN prices from 1998 to 2000, provide little indication of large cyclical price movements. In both cases there was an initial surge in prices before the period covered by the graphs. This was followed by a steady decline from 1999 onwards with only minor increases near the reporting cut-off date. Profiteering can therefore only have been sporadic rather than on the concerted basis suggested by compliance schemes. Subsequent analysis has shown that producers' fears were fuelled by large price increases during the early stages of the PRN system, which were subsequently attributed to the actions of VALPAK, Britain's biggest compliance scheme. When PRN trading began, it was anxious to secure

Figure 4.1 PRN prices (£'s per tonne), excluding plastics

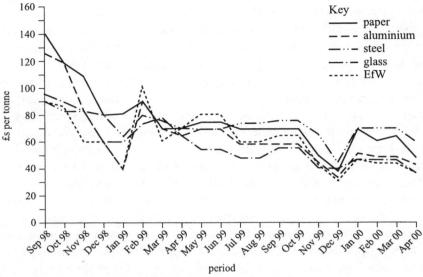

Source: Materials Recycling Week, monthly data (September 1998-April 2000)

Figure 4.2 PRN prices for plastics (£'s per tonne)

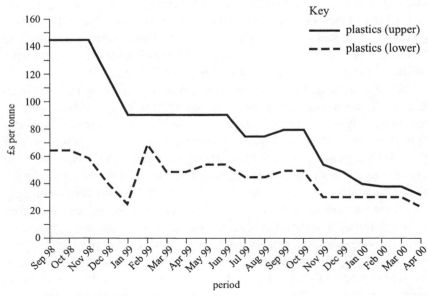

Source: Materials Recycling Week, monthly data (September 1998-April 2000)

certificates for its large membership and bought hefty quantities of PRNs early in the annual cycle, usually via fixed contracts with major reprocessors (Bailey, 1999b). Because no-one could foretell how much packaging waste would be reclaimed in the first year, this pushed up PRN prices until more uncommitted materials became available later in the year[13]. In order to stabilise the market's annual cycle more compliance schemes have taken to signing annual contracts for the supply of PRNs.

It therefore appears that a combination of competition amongst reprocessors and the purchasing power of compliance schemes has prevented blatant hoarding. Even in sectors where competition is less intense, such as the steel and aluminium sectors – which are dominated by major players and market entry is restricted by high start-up costs – reprocessors claim to have adopted a 'benign' stewardship role, using PRN funds to stabilise the sales price of recycled materials. According to interviewees from these sectors, PRN revenue is only a minor income stream so there are few commercial or publicity advantages to be gained from undermining the market. The government nonetheless accepted that the sale of PRNs to speculators distorted markets and banned the practice (DETR, 1998b). It also sought to give producers that collect their own packaging waste first refusal on the PRNs issued for these materials.

The second and more damning criticism was that accredited reprocessors were not using PRN revenue for infrastructure development but were simply pocketing it as windfall profit whilst making negligible efforts to expand recycling or support material prices. As the system placed no statutory obligations on reprocessors aside from the requirement to submit PRN returns to the Agency, it created no *binding* guarantee of increased investment (ACP, 1998). In many cases this has caused considerable disenchantment amongst producers and compliance schemes, who thought they were meant to be contributing, however grudgingly, towards more sustainable waste management. As the expansion of reprocessing and collection capacity was one of the fundamental challenges faced by the British packaging waste system, the general consensus was that this aspect of the policy was critically flawed.

The DETR's review of the reprocessing industry added further credence to this indictment when it confirmed that several sectors, particularly plastics and EfW incineration, were lagging behind EU targets (DETR, 1999a). Whilst the report concluded that the PRN system and broader economic factors had both contributed to this deficit, one compliance scheme was more categorical in its condemnation:

The glass, aluminium and steel areas have been a success simply because reprocessors have not needed to invest one jot of new capacity to reprocess recovered material. Those sectors requiring fundamental investment (because the existing infrastructure was incapable of operating on reclaimed rather than virgin input resources) such as paper, card and plastic have demonstrated a complete inability to prove where the funds have gone – and we strongly suspect that those funding flows have propped up balance sheets or profit and loss at a time of falling global commodity prices.

Reprocessors countered these accusations by arguing that the way in which PRN trading was organised was not conducive to long-term investment (DETR, 1998b). They pointed out that the annual compliance deadline, fluctuations in PRN prices, and uncertainties over the future of the Regulations have dissuaded most compliance schemes from entering into long-term contracts for PRNs (typically one year is the maximum). As the pay back for a new reprocessing plant may be several years, reprocessors felt there were insufficient guarantees to warrant major investments, especially in a climate of variable international materials prices. Two respondents from the plastics reprocessing sector made similar points:

> There is a potential conflict of interests between the compliance schemes and the reprocessors. Compliance schemes wish to obtain PRNs at the most economic rate and to remain flexible in what is a rapidly changing system. They are therefore reluctant to enter into longer term purchasing contracts. This can potentially work against reprocessors on both fronts by failing to create the conditions for a sustainable and expanding reprocessing market.

> It is difficult to invest in new facilities when long term guarantee of price (or demand) for PRNs cannot be given.

Reprocessors also complained that expansion in the PRN market was being stunted by the large number of producers failing to register with the Environment Agency and joined calls for tougher legal action in high profile cases (O'Doherty and Bailey, 2000). Not all reprocessors shared this pessimism, however, and a significant number indicated that they thought the scheme was working effectively and that they were using PRN monies for the purposes intended by the Environment Agency:

> All PRN revenue is used to promote recycling in the UK either by supporting price, or developing infrastructure or promoting awareness

campaigns. If there is a surplus of recyclate the revenue can be used to support prices, conversely if there is a surplus of manufactured recycled end product then the revenue can be used to support lower sales prices to stimulate end markets. (aluminium)

The revenue from PRNs enables us to buy waste which would not otherwise be viable commercially to pay for ... and to help finance expansion schemes. (plastics)

It is our view that the current PRN situation is such that it provides economic justification for the installation of Energy from Waste facilities without which such projects would not proceed. (incineration)

Table 4.4 presents an overview of reprocessors' opinions towards the PRN system, grouped by sector. The strongest negative response came from the plastics sector, providing a further indication of its disappointment at the effects of PRN trading on the profitability of plastics recycling. Whilst paper reprocessors were more pessimistic than one might expect, the survey was conducted at a time when PRN prices for paper had fallen to low, and in some cases negative, levels[14]. Conversely, the apparent optimism amongst incinerator operators reflects the fact that many are receiving the benefit of PRN cash even if it is not enough to encourage significant expansion in the industry. In fact, the overall the attitude of reprocessors towards the PRN system seems to be determined more by whether it has made their businesses more profitable than whether it has allowed expansion across the sector to produce wider social and environmental gains.

Table 4.4   Reprocessor opinions towards the PRN System

Activity	Number of respondents	Negative	Percentage of firms Neutral	Positive
Paper	13	53.8	46.2	0.0
Glass	3	0.0	33.3	66.7
Aluminium	2	0.0	50.0	50.0
Steel	4	0.0	25.0	75.0
Plastics	17	58.8	23.5	17.7
Incineration	7	14.2	42.9	42.9

Reactions to the PRN system have therefore been mixed. On the one hand, some reprocessors have welcomed the introduction of a fiscal mechanism that supports their industry. On the other, there is considerable disenchantment amongst both buyers and sellers on the PRN market and the balance of evidence suggests that weaknesses in the scheme are a key reason factor behind the slow development of recycling infrastructure in the UK. In short, the combination of recycling targets for producers and market inducements has not been enough to persuade a sufficiently large number of reprocessors to invest in new facilities and, as a result, exports for reprocessing have risen (ACP, 2001). It might even be possible to argue that similar or superior results could have been achieved using legislation and a straightforward recycling charge without the experiment in tradable certificates. Though this counterfactual argument is difficult to verify, the DSD expanded its recycling network using precisely this methodology.

Two principal factors explain why PRN trading has not brought about sustained growth in recycling activity. First, the material markets where expansion was most needed tended to remain unprofitable despite the injection of PRN cash. This has been caused in part by fluctuations in international commodity prices but compounded by a general slide in PRN prices and the absence of a clear mechanism specifying how reprocessors should spend PRN revenue. Transitional subsidies for problem sectors might help to overcome some of these problems but these would require direct intervention in PRN pricing or a commitment to public expenditure funded from another areas. It might also create inter-sectoral conflicts if subsidies were provided from sectors that have achieved or exceeded their reprocessing targets. Overall, the government has sought to avoid cross-subsidisation in order to maintain equity between sectors.

Whilst capricious international prices are an unavoidable fact of life, the experimental nature of the PRN system has itself caused many problems. For example, most permit schemes impose a *limit* on the production of regulated pollutants but the PRN system seeks to *promote* a range of environmentally beneficial activities. Generally speaking it is more straightforward for the authorities to control the former than the latter. Similarly, it is unusual for permit schemes to stipulate how proceeds from trading should be spent. Although governments often expect firms that sell permits to invest in abatement technologies, the environmental efficacy of the scheme does not necessarily depend on this because restrictions have already been imposed on the pollutant. Pollution limits can also be altered if the authorities see fit through the release or recall of permits. To achieve the same effect the UK government would need to raise recycling targets on a 'dose-response' basis until investment began to increase significantly, though options for protracted experimentation are constrained by the Directive's implementation deadline. Finally, trading is usually restricted to the sector affected by the scheme and recognised brokers (Hahn, 1993). The innovation of market trading between waste originators and reprocessors, which appeared to offer a range of synergies, has not thrived because the government did not fully anticipate the impact of external commercial pressures on the behaviour of reprocessors. Equally, the Conservative administration, which instigated PRN trading, was naturally more inclined to see the benefits of the free market and therefore failed to establish sufficient controls on trading and investment.

The present administration has been generally cautious towards reforming the PRN scheme despite deriding the Regulations when in opposition (House of Commons, 1997). The DETR's 1999 review exhorted industry to increase recycling but suggested few major alterations to the PRN scheme. Tentative steps towards reform instead included: barring speculators from PRN trading, new reporting requirements for reprocessors to detail the proportion of PRN income being earmarked for recycling infrastructure, and guidelines for de-registering reprocessors that failed to abide by the spirit of the scheme (DETR, 1998b). However, in order to maintain the system's onus on market-led recycling, the government stopped short of more draconian measures, such as prescribing minimum investment requirements. In recent interviews the Agency claimed that a combination of increased recovery targets and the additional controls were beginning to be reflected in increased PRN prices and infrastructure investment (MRW, 2000). Similar views were expressed by the DETR and in 1999 it decided to increase the 2001 recovery target to 56 per cent in order to provide a boost to all elements of the PRN system (DETR, 1999d). The DETR felt this move would force producers to collect more waste, causing PRN prices to rise with the accelerated demand for compliance certificates. This, in turn, would give reprocessors a greater financial incentive to invest in collection and recycling infrastructure.

There were also indications in 2000 that the government was beginning to lose patience with market-based recycling. Around this time the Environment Minister, Michael Meacher, considered proposals to cut reprocessors out of the PRN system altogether, leaving them to finance the acquisition and marketing of recovered waste through normal channels without PRN subsidies. This proposal seemingly reaffirmed that collection had become a higher priority than reprocessing, as it also included plans to send PRNs to local authorities for packaging they collect, allowing them to raise funds independently for expanding collection services. This view was even supported by one respondent from the reprocessing sector:

We believe the Government should control the flow of PRNs by purchasing all PRNs on January 1st and selling them to obligated companies at a fixed price. This would take unnecessary fluctuations out of the market and allow UK companies in both sectors to plan and budget their expenditure in this area.

Such comments suggest that although reprocessors have frequently been portrayed as the villains of the PRN system, they have also experienced genuine difficulties operating within its present structure. It should also be noted that recent reports are more optimistic about the state of UK recycling. In 2001, the majority of compliance schemes reported to the ACP that they were confident of meeting their share of the recovery obligation, whilst materials organisations noted that a combination of new facilities and PERNs had overcome the worst reprocessing bottlenecks (ACP, 2001; Department for Environment, Food and Rural Affairs (DEFRA), 2001). The ACP nonetheless opined that many issues still needed to be resolved if the UK was to guarantee compliance with EU targets in the long run. Of these, waste collection and improved co-ordination between compliance schemes, reprocessors and local authorities were highlighted as the most immediate priorities. The ACP also noted that it preferred the development of more contractual relationships within the recycling chain to the exclusion of reprocessors from the PRN scheme.

In terms of overall diagnosis, it appears that a combination of revised targets, reforms to monitoring and enforcement procedures, and market-led PRN trading has eventually produced the increases in recovery and recycling sought by the government. However, many weaknesses in the design of PRN trading have been exposed and the environmental value of increasing recycling through the waste exports remains particularly questionable. The majority of problems have nonetheless been caused by regulatory deficits in areas where ground rules were needed to control the actions of market players rather than the outright failure of market-led recycling. In keeping with this conclusion, most of the remedies implemented or being considered are of a command-and-control nature.

4.5   Conclusions

The Packaging Waste Directive presented the EU member states with a multitude of political and practical challenges to which the British and German governments have responded by deploying a variety of NEPIs. Negotiated agreements were used to establish producer responsibilities and to apportion targets between sectors of the packaging chain, cost-covering charges were introduced to finance expansion in reprocessing and collection capacity, and a tradable-permit scheme was initiated in the UK to facilitate transactions and mitigate producers' compliance costs. In both countries, however, legislation has been the main instrument for transposing EU environmental requirements into national law and NEPIs have been employed principally to manage aspects of implementation that could not be addressed using command-and-control techniques.

By examining implementation problems experienced in the two countries, the Packaging Waste Directive provides some useful lessons on the general strengths and limitations of these NEPIs. In particular, it has highlighted the difficulties arising from excessive reliance on market forces as a medium for implementing environmental policies. Where there is a natural coincidence between the interests of the market and environmental protection, or where one can be induced through adjustments to the price mechanism, MBMs can make a substantial contribution towards reducing the cost of environmental policies. Where they do not, the market will tend to prioritise economic efficiency at the expense of environmental effectiveness and secondary regulation may be required. The PRN system initially failed to induce adequate expenditure on new recycling infrastructure because there were few financial incentives and no mandatory requirements imposed on individual reprocessors. Considering the experimental nature of the PRN system some iterative adjustments were only to be expected but, by the same token, the government's failure to regulate the scheme appropriately at an early stage led to considerable industry disenchantment with the market-led approach to environmental policy. Conversely, over-reliance on command-and-control can also produce negative consequences by making the implementation of environmental policies more expensive. Whilst an economic instrument was introduced in Germany to promote infrastructure investment, it has been circumscribed by regulation (high recycling and re-use quotas) and efforts to make the Green Dot more cost effective have been extremely belated. This has produced a system that is significantly more expensive than the PRN scheme, both in absolute and relative terms, which may in turn increase industry antagonism towards future environmental policies of this type.

Finally, the importance of national political preferences in determining the design and deployment of NEPIs has been re-affirmed. Whilst the PRN system is generally a more faithful interpretation of economic theory than the DSD, national control over the design of NEPIs has nonetheless increased the scope for divergence in national policies. This has resulted in wide variations in *de facto* environmental outcomes achieved in the two member states. Whilst some amount of controlled diversity may not be an immediate concern for the Commission provided minimum requirements are met, it may have implications for the harmonisation of national policies, the functioning of the internal market, and progress towards sustainable development. The balancing act for the Commission is to maintain a structure of environmental policy that allows flexible implementation in accordance with subsidiarity and the environmental capabilities of each member state, whilst ensuring that *flexibility* does not spill over into *fragmentation* of the EU environmental programme. This issue is examined further in Chapter six. Chapter five now examines the potential of environmental charges in the UK and Germany to act as an incentive instrument for encouraging more sustainable waste management amongst businesses affected by producer responsibility.

Notes

[1] For reviews of the German Packaging Ordinance, see London and Llamas (1994), Michaelis (1995), Simonsson (1995), Waite (1995), Whiston and Glachant (1996), Staudt (1997), Flanderka (1998), Eichstädt *et al.* (1999) and Haverland (1999, 2000a). For analysis of the UK Packaging Regulations, see Bailey (1999a; 1999b; 2000), Haverland (1999; 2000a), Nunan (1999) and O'Doherty and Bailey (2000).

[2] Data supplied by the Packaging Federation.

[3] Carbon leakage occurs where industries in states that have ratified the Kyoto Protocol reduce their greenhouse-gas emissions but this contraction is offset by increases in countries that are exempted from the Protocol or which have refused to sign. This effect is strongest in energy-intensive sectors and has prompted lobbying for reduced carbon taxes for these industries to protect their competitive position. In exchange, sectors may be asked to make voluntary commitments to achieve higher environmental standards.

[4] The first successful prosecution for breach of the Packaging Regulations took place in May 1999, where the company in question was fined £150 and ordered to pay £1,424 in costs (ENDS, 1999a). A second successful prosecution was secured one month later but the fine imposed (£750) was equivalent to the cost of registering with the Environment Agency and therefore its deterrent effect was highly questionable (ENDS, 1999b). At the time of writing the average fine is £1,500, with sentences ranging from an unconditional discharge to a maximum fine of £10,000 (ENDS, 2000e).

[5] This compares with £1.37 billion spent on recycling services out of a total revenue of £1.46 billion in 1998, suggesting that hypothecation levels consistently achieve 94 per cent (DSD, 1999a, 1999b). The reduction in sterling expenditure is the result of a combination of reduced DSD income and exchange rate fluctuations between 1998 and 2000.

[6] These figures are based on the total number of PRNs traded in 1998 and average PRN prices over the period. They are therefore only an approximation of PRN revenue available, as trading and prices vary from month to month but full trading data are only published annually.

[7] In *A Way with Waste*, the government estimated that if recycling of municipal waste remained around 25 percent, one third this waste stream would require incineration in order to fulfil the requirements of the EU Landfill Directive. This directive establishes, *inter alia*, quantified targets for reducing the proportion of biodegradable waste going to landfill. To meet this objective, between 28 and 130 new incinerators may need to be commissioned, depending on the capacity in each incinerator. Packaging is, of course, only a minor component of municipal

waste (DETR, 1999c).

[8] In one example, an interviewee from the plastics sector claimed that the average cost of recycling one tonne of plastics in 1998 was £270 compared with a PRN price of £140. The company was therefore still strongly dependent on end sales to recoup its costs but the unpredictability of the plastics market meant this was only intermittently achieved. PRN prices for plastics have subsequently fallen to an average of £23-32 per tonne.

[9] Raymond Communications reported in its October 2001 on-line bulletin that the DSD has come under increasing attack in Germany and from the Commission for allegedly monopolistic tendencies. Several local governments and retail chains are currently trying to set up their own schemes in direct competition to the DSD.

[10] Latest data from the Environment Agency indicate that, in 2000, 709,000 tonnes of packaging waste was collected from households out of a total of 3,846,743 reprocessed. This represents 18.4 per cent of total collection. However, the Agency estimates that there is only 1,539,438 tonnes of material annually left in commercial-industrial waste streams compared with 4,428,500 tonnes of domestic material. If, as the ACP predicts, packaging waste increases from 9.3 million tonnes in 2001 to 9.62 million tonnes by 2006 and a 60 per cent recycling target is agreed as part of revisions to the Directive, between 1.46 million and 1.59 million tonnes of household waste will need to be collected (ACP, 2001). However, the real figure is likely to be much higher because it is not feasible to collect all commercial-industrial waste. Furthermore, when considering the role of local authorities in the recycling chain, it should be remembered that they collect a significant proportion of commercial-industrial waste. It is also worth noting that calculating waste arisings is an extremely imprecise science. The ACP working parties admit as much in their reports and their figures are routinely re-examined (ACP, 2001). This accounts for discrepancies between these data and those in Table 4.3, which utilises DETR 1999 data, the only report published to date that compares reprocessing capacity against projected waste volumes in the UK (DETR, 1999a).

[11] Interestingly, this idea has recently re-surfaced in response to pressures caused by the EU's Landfill Directive. A variant of direct charging was piloted in Denmark in the 1990s, where weight-based waste fees were introduced for non-compostable household waste. At the start of the scheme, waste was weighed on collection and households were charged at the same rate as the waste collection component of municipal taxes. As the charge per unit of waste (by weight) did not change, households had a positive financial incentive to reduce waste and taxes through increased segregation and recycling. The result was 15-20 per cent reduction in the weight of recyclable refuse collected as waste (Barrett *et al.* (1997:112).

[12] The term 'banking' refers to the storage of excess permits for future use, whereas 'borrowing' involves using permits ahead of their allocated date on the understanding that improvements in abatement technologies and techniques will occur during the period (Grubb *et al.*, 1999). With recycling, a banking scheme for PRNs would allow producers to exceed their recovery target in one year in return for a reduced target in future years. With borrowing a producer would be excused a deficit in one year provided the shortfall is redeemed at a later date. For such schemes to operate in an effective and equitable manner, the period over which banking and borrowing is permitted must be specified and legally enforceable.

[13] This meant VALPAK purchased PRNs at a comparatively high price, making it less attractive to potential members. VALPAK attempted to counteract this by demanding that reprocessors offer a rebate in future years should PRN prices fall later in the year. The Department of Trade and Industry saw this as a restrictive practice, however, and blocked the move (MRW, 1998a).

[14] During 1999 there were unconfirmed reports that paper reprocessors were actually charging waste collection firms for delivering waste paper to their premises because paper PRNs commanded no value.

Chapter 5

Corporate Responses to Environmental Taxes and Charges

5.1   Introduction

It was noted in the previous chapter that the UK and German governments introduced economic instruments for packaging waste principally to raise revenue from producers for investment in new recovery and recycling infrastructure. According to the typology developed by Ekins (1999), therefore, neither the Green Dot nor the PRN scheme were conceived as incentive charges and should not be judged exclusively as such. It is particularly important to remember that neither instrument was designed to restrict the consumption of packaging and packaged goods, as such an approach could compound the economic impact of the DSD and detract from the British government's intention to implement the Packaging Waste Directive in a cost-effective manner. Hypothecated charges nonetheless have the potential to affect the production and management of packaging waste in ways that may be considered desirable by national policy-makers:

- Companies may be encouraged to 'light-weight' packaging where savings on environmental charges and material requirements can be gained from reducing excess packaging;
- The re-use of packaging materials may be encouraged. The UK Regulations, for example, allow packaging which is re-used three or more times to be excluded from producer responsibility returns. By switching to re-useable packaging, therefore, businesses can reduce the financial liability of recycling charges;
- Firms in the UK that fall close to the Regulations' 50 tonne annual threshold for packaging consumption have an incentive to reduce their packaging throughput to exempt themselves from producer responsibility. Economic instruments may therefore encourage some reduction in packaging consumption even when set at a relatively low rate. If sufficient companies follow this course of action, however, this may also marginally reduce national recycling rates[1].

Both the British and German economic instruments for packaging waste could therefore conceivably produce incidental incentive effects to complement their main cost-covering function and such actions would also promote the Directive's 'Essential Objectives' of reducing the production of packaging waste and increasing its re-use. In fact, no other policy instruments aside from the Green Dot, PRNs and guidelines on the design of packaging have been clearly defined in either country to implement these objectives (Eichstädt *et al.*, 1999)[2]. The purpose of this chapter is to explore the extent to which hypothecated charges have produced financial incentives for packaging producers to engage in more sustainable waste management. The discussion begins by reviewing national trends in packaging production then examines producers' responses to recovery and recycling charges. Following this, the link between environmental charges and producer policies is examined in more detail and critical factors determining the incentive effect of economic instruments are discussed. The chapter concludes by offering a brief commentary on options for the use of economic instruments in environmental policy.

5.2   National Indicators of Packaging Production

A first indication that recycling charges have produced a discernible effect on packaging producers comes from national government data. The UK and German environmental ministries each maintain records of the amount of packaging waste entering the waste stream using data provided by materials organisations. The data for Germany provide a clear indication that an overall decline in packaging consumption has occurred. The DSD reported that per capita packaging consumption in Germany fell from 94.7 kilogrammes in 1991 to 82.0 kilogrammes in 1998, a reduction of 13.4 per cent (DSD, 1998a; 2000). The DETR's projections for the UK provide a more uneven picture of trends in packaging consumption, however, with sizeable variations between the different packaging materials (Table 5.1). In 1998, DETR forecast that the quantity of glass, steel and wood entering the waste stream would not increase before 2001, though aluminium waste was predicted to grow by 0.9 per cent per annum and plastic waste was anticipated to rise four per cent each year between 1998 and 2001 (DETR, 1999a). More recent data suggest some of these predictions were over-pessimistic, with less growth than expected in plastics but an increase in paper packaging waste (ACP, 2001). The DETR and the ACP also acknowledge that the amount of wood packaging entering the waste stream has proved difficult to quantify because of the influence of wooden pallets. Total packaging consumption nonetheless rose

Table 5.1   UK recovery and recycling 2001 (thousands tonnes)

Material	1998 DETR estimates	Annual growth factor (%)	2000 predicted	2000 reported	2001 revised estimate
Paper	4000	2.5	4202	3855	3855
Glass	2200	0.0	2200	2155	2200
Aluminium	109	0.9	110	110	120
Steel	735	0.0	735	750	750
Plastics	1700	4.0	1839	1600	1679
Total exc. wood & other	8744		9086	8470	8604
Wood	1300	0.0	1300	670	670
Other	200	0.0	200	40	40
Total	10244		10586	9180	9314

Source: DETR (1999a: 10-11):(DEFRA 2001:5)

slightly between 2000 and 2001, though the overall trend between 1998 and 2001 suggested static or slowly decreasing consumption. At the time of writing the ACP is predicting a small incremental growth in packaging waste over the period 2001-2006 (ACP, 2001).

When interpreting these data, one of the key factors to take into account is the scale of charges imposed on packaging materials in the UK and Germany (see Figure 3.6). As noted in Chapter three, German Green Dot fees are significantly higher than prevailing PRN charges for all types of packaging waste, with the greatest differentials occurring in the levies on plastics and aluminium. This implies that higher recovery charges – or other market or regulatory factors – in Germany have created clear incentives for packaging reduction whereas vagaries in data presentation and lower charges in the UK mean that similar trends are difficult to detect. DSD publicity appears to confirm this by highlighting that noteworthy reductions in packaging consumption are occurring in Germany, with numerous accounts of companies embracing the practice of 'light weighting' (DSD, 1998a).

However, it is important that the incentive effect of packaging-waste charges is not overstated on the basis of this information. First, the DSD only deals with sales packaging and figures for secondary and transport packaging are not available from the German environment ministry. It is therefore possible that reliance on DSD returns yields a biased picture of packaging trends in Germany. The effects of partial data are best illustrated by comparison with the UK, where numerous companies have begun substituting sales packaging with transport packaging because the latter can be more readily recovered by producers during transit through the supply chain (ENDS, 1996). In some cases, this has been counterproductive to the government's recycling strategy since the escalation in transport packaging has contributed towards increased waste arisings whilst nonetheless facilitating individual firms' compliance with the Regulations. Since the provisions in the Packaging Ordinance obliging German retailers to provide on-site collection bins for unwanted sales packaging were waived following the foundation of the Dual System, some German firms may have adopted similar ploys. However, although the financial impact of the Green Dot would diminish as sales packaging reduced, this would be offset by other recovery costs because companies are obliged to make independent arrangements for non-sales packaging. It is nonetheless important to bear in mind that DSD data on packaging consumption represent only part of the wider picture.

The second reason why it is risky to assume a causal relationship between waste charges and packaging consumption from these data is that a range of regulatory tools were introduced in both countries to implement the Packaging Waste Directive, including statutory instruments, indicative 'essential requirements', voluntary agreements and informational devices. One should not presume, therefore, that the reductions in packaging consumption observed are entirely attributable to economic instruments. Indeed, this is extremely unlikely and further testing is required to isolate the effects of environmental charges from those produced by other policy instruments.

5.3   Producer Perceptions of Economic Instruments

In order to explore the effects of economic instruments on the production and management of packaging waste further, surveys were conducted with UK and German packaging producers. Their aim was to identify whether companies with a significant stake in packaging or packaged goods felt that economic instruments had encouraged major changes in waste management. To achieve this aim, the following key variables were measured:

1. Actions taken by obligated companies in response to national packaging policies; have producers introduced defined waste management plans to fulfil the objectives of the Packaging Waste Directive? The types of waste management assessed in the survey were the following:
   - Source reduction
   - Re-use
   - Collection of packaging waste for recovery and recycling

- Purchase of packaging manufactured using recyclate (development of end markets);
2. Green Dot or PRN charges incurred;
3. Perceptions towards the environmental and economic efficacy of packaging-waste charges;
4. Opinions on other policy instruments introduced to expedite the Packaging Waste Directive, such as statutory instruments and voluntary agreements;
5. Attitudes towards alternative policy options. For the purpose of the surveys, an inventory of policy alternatives was compiled, all of which, at one time or another, have been considered by either the UK or German governments.

Two notable omissions from these research questions require justification. First, although the waste management options examined correspond to the main 'Essential Objectives' of the Directive, no attempt was made to quantify respondents' performance against statutory recovery and recycling targets. Pilot studies showed that few companies were prepared to divulge such legally sensitive information, even under the cloak of respondent anonymity. Questions of this nature were therefore restricted to general enquiries as to whether businesses would meet or exceed their legal obligations and the inclusion of collection as an indicator which could be used to infer firms' direct involvement in the recovery and recycling chain. Second, only PRN or Green Dot charges were considered in the surveys although it is readily acknowledged that firms may incur indirect costs in complying with the Directive. Indirect compliance costs may include, *inter alia*, data collection, staff training, modifications to information technology, and consultants' fees, but were deliberately omitted because they tend to be loosely defined and specific to each firm and would therefore hamper reliable comparisons. Additionally, although indirect costs may contribute significantly to overall compliance costs, they stem mainly from legal obligations rather than economic instruments *per se* and are therefore not directly germane to the analysis. Finally, many companies do not accurately monitor indirect costs and, consequently, any data of this type are likely to be unreliable.

In total, 900 businesses from each country were surveyed using stratified random sampling of registers provided by the UK Environment Agency and German corporate directories. Replies were received from 469 British firms (52.1 per cent) and 309 German companies (34.3 per cent). Filter questions were used to eliminate businesses not affected by national packaging regulations: this reduced the number of useful responses for the UK to 450 (50 per cent) and 236 (26.2 per cent) for Germany. Company profile data were also requested in respect of turnover, number of employees and sector to establish whether responses to environmental charges were affected by particular corporate characteristics. However, the influence of these factors was found to be negligible and are therefore not reported in detail.

5.4   Corporate Waste Management

The first element of the surveys identified the percentage of British and German firms engaged in each form of waste management promoted by the Packaging Waste Directive, excluding the mandatory recovery and recycling targets (Table 5.2). For all indicators, the number of German companies with defined waste management plans far exceeded those in Britain. This division was especially pronounced for waste prevention, with 57.1 per cent of German respondents having instituted source reduction plans for packaging waste compared with 12.7 per cent of British firms. A sizeable portion of German industry therefore appears to have adopted strategies that seek to prevent packaging waste from occurring in the first place – in addition to their involvement in recycling – whereas relatively few British businesses have embraced a similarly preventative approach. The surveys revealed less (although still significant) discrepancy in relation to the re-use of packaging waste, which might initially appear surprising given the re-use provisions in the Packaging Ordinance. However, this is explained by the inclusion of wood packaging and, in particular, wooden pallets in the UK Regulations. Wood was the most common form of packaging re-use reported by British businesses, reflecting the popularity of pallet-control systems such as GKN Blue Chep[3].

The one area where there was no significant difference between the waste management strategies of UK and German firms was in respect of collection,

Table 5.2   Waste management actions, British and German companies

Action	Per cent of firms engaged in waste management activity	
	UK	Germany
Reduction in consumption	12.7	57.1
Re-use	23.8	52.4
Collection from business premises	64.5	85.4
Collection from customers	13.4	18.4
Purchase of recycled packaging	24.5	53.2

Source: Bailey (2002: 244) *(reproduced courtesy of Elsevier Science)*

where most respondents reported routinely collecting and separating waste arising at company sites but have largely avoided reclaiming used packaging from customers or consumers. Both trends are in fact indicative of the structure of the two national recycling systems, in that local-authority or DSD contractors undertake the majority of post-consumer waste collection, particularly for sales packaging. Waste collection at business premises usually involves secondary or transport packaging where firms can minimise recycling costs by recovering high volume and homogenous commercial-industrial waste which is close at hand. However, the high cost of recovering waste through supply chains has dissuaded most firms from engaging in post-consumer collection.

It must be borne in mind, however, that the mere event of a firm collecting, reducing, or re-using particular types of packaging waste provides little indication of the scale of its involvement in these activities. For example, a business that collects a small percentage of waste office paper cannot reasonably be compared to one that collects large volumes of residual materials from a production or transport process. The surveys therefore also sought to assess the extent of corporate involvement in each activity by measuring waste management targets as a percentage of total packaging consumption for 1998 and predicted performance in 2001 (Table 5.3).

Table 5.3   Waste management performance and targets

Target	Period	Mean target 1998-2001 (per cent of packaging)	
		UK	Germany
Reduction in consumption	1998-2001	2.0	10.8
Re-use	1998	10.5	24.3
	2001	15.8	32.1
Collection (combined)	1998	21.1	40.3
	2001	27.5	45.6
Purchase recycled packaging	1998	14.4	37.2
	2001	20.5	47.4

Source: Bailey (2002: 245) *(reproduced courtesy of Elsevier Science)*

From these data it is again evident that German businesses have engaged more extensively with the idea of waste management (as opposed to merely disposal) than their British counterparts. For example, the mean source reduction target for packaging waste amongst British respondents was two per cent between 1998 and 2001, compared with targets in excess of ten per cent for German firms over the same period, and similar differentials were found for re-use, collection and the purchase of packaging made from recycled materials. At the same time, the data provide indications that British firms are closing the gap in some areas. Although British respondents predicted that packaging re-use would rise by 5.3 per cent in the period 1998-2001 compared with 7.8 per cent amongst German companies, waste collection is anticipated to increase by 6.4 per cent in the UK as opposed to 5.3 per cent in Germany. Finally, the percentage of packaging made from recycled materials used by respondent businesses is expected to rise by 5.9 per cent in the UK, against 10.2 per cent in Germany. When compared with 1998 performance, however, British industry has secured greater relative improvements in all areas of waste management, chiefly because its recycling networks are expanding from a low starting point, whilst those in Germany are relatively well developed and further expansion may be hampered by diminishing returns. The general impression is nonetheless that British industry trails Germany by a considerable distance in packaging waste management even though the UK Regulations are beginning to yield tangible results. A similar trend of 'catching up' is also evident in relation to the Directive's statutory targets, with recent DEFRA surveys reporting a three to four per cent annual increase in recycling between 1998 and 2000 (DEFRA, 2001).

In terms of packaging materials, the greatest changes in waste management have occurred in the paper and plastics sectors, with the majority of firms surveyed routinely collecting waste paper packaging and buying packaging made from recycled paper or card (Figures 5.1 and 5.2). Again, these trends reflect the well-developed state of paper recycling in Britain and Germany. However, there is also a clear trend towards the reduction of paper and card packaging in Germany, with over 55.8 per cent of respondents reporting that source reduction measures had been instigated. A major distinction exists for plastics, however, with 40.1 per cent of German companies reducing consumption of plastic packaging compared with 9.3 per cent of British businesses. The re-use of plastics packaging has also increased in both countries with, for example, the introduction of tote bins to replace cardboard boxes for small, low-value consumer goods[4]. The main driver behind the reduction and re-use of plastic packaging has been the higher costs of recycling, which are reflected in Green Dot and PRN prices. By contrast, glass,

Figure 5.1 Packaging waste management by material, UK

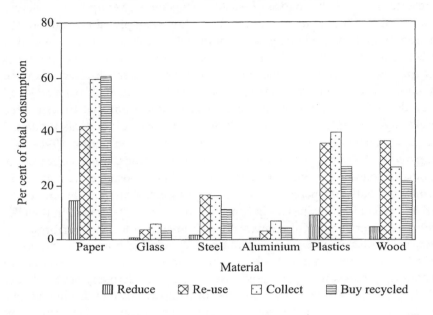

Figure 5.2 Packaging waste management by material, Germany

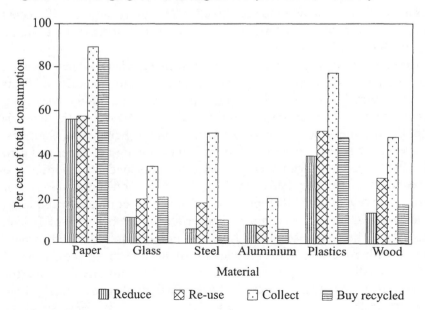

aluminium, and steel are mainly recycled rather than reduced or re-used because recycling costs are comparatively low and the majority of infrastructure is already in place for both commercial-industrial and post-consumer collection.

In addition to taking these actions to promote more sustainable waste management, both respondent groups generally thought that current government policies were environmentally effective. 71.1 per cent of German firms felt the Packaging Ordinance and DSD had changed the way industry treated packaging waste and 61.1 per cent felt they were effective in promoting reduction, re-use, and recycling. 53 per cent of British respondents also recognised that the Packaging Regulations had provoked positive changes in corporate waste management and 51.5 per cent predicted that waste management would become more environmentally sustainable as a result. However, German companies were more optimistic than their UK counterparts about their ability to meet the requirements of the Directive and expressed greater confidence that the environmental benefits of the Directive justified the costs incurred by industry (Table 5.4). It is interesting to note that over 70 per cent of German companies felt they would exceed the requirements of the Directive, presumably because the German legislation contains additional targets. Whilst UK firms accepted the environmental logic of the Directive and the need for legislation to remove trade barriers, they were more suspicious about the implementation methods employed by the government. Whilst this can partly be attributed to the fact that, at the time the survey was conducted,

Table 5.4   Corporate opinions on national packaging regulations

Proposition	Per cent of companies agreeing with proposition	
	UK	Germany
The company will only achieve the Directive's minimum statuory requirements	53.7	40.5
The company will exceed the Directive's targets	23.0	70.7
National regulations will achieve cost-effective environmental protection	16.9	43.3
Cost of national regulations to industry are justified by the environmental benefits gained	42.8	60.7

the Regulations were a relatively recent imposition and firms were still acclimatising to their requirements, confidence in government policy had also been badly dented by the PRN system's early problems. By contrast, German industry has now operated within the strictures of the Dual System for eleven years and has therefore had time to adapt and to see the environmental benefits of recycling.

Despite the prevailing mood of optimism amongst German companies, only a minority of businesses in either country felt that the targets established by the EU and national governments were reasonable and just 24.5 per cent of German firms and 24.3 per cent of UK respondents opposed the lowering of recovery targets. There was also a strong sentiment that the numerous derogations and opt-up clauses in the Directive should be removed to eliminate any lingering trade restrictions. In fact, over 85 per cent of companies in both countries supported this view. Interestingly, some large UK firms opposed the idea of greater harmonisation on the grounds that it might lead to more stringent targets if, for example, German standards were adopted across the EU. This demonstrates the more nuanced policy awareness of larger companies, which have more experience of monitoring EU environmental policies, as well as their desire to preserve a competitive advantage in European markets through the maintenance of legislative differentials. Nonetheless, for different reasons, neither respondent group considered the current legal framework to be satisfactory.

In summary, the Packaging Ordinance and its associated implementing mechanisms have produced major changes in the way that German industry perceives and manages packaging waste. Although the DSD has failed to achieve national re-use quotas for beverage containers, most other objectives and EU requirements have been achieved. Equally, these changes have occurred despite continued uneasiness about the cost of the Packaging Ordinance and the consequences of this for the competitiveness of German industry. Their acceptance that the Packaging Ordinance is environmentally justified and that industry is capable of exceeding current targets are particular significant, as they demonstrate an evolution in business responses to the issue of waste management. German firms appear to have moved beyond resistance to, or passive compliance with, legal requirements to embrace a more active and pre-emptive approach to waste management (Welford and Prescott, 1994). The increases in waste prevention and re-use predicted by UK respondents for the period 1998-2001 suggest that British industry may be undergoing a similar transition and this is particularly significant considering that the Regulations have been in force for a comparatively short time. One of the most interesting trends to emerge, however, is the different perceptions towards the cost-effectiveness of national packaging policies, where German companies appear to have accepted that the environmental benefits of recycling justify the costs of the DSD but UK firms continue to see compliance costs as a major bone of contention. Having reviewed the main changes in waste management that have occurred in the UK and Germany, the discussion now returns to examine the effect of economic instruments on corporate waste strategies.

5.5   Economic Instruments and Packaging-Waste Management

Establishing direct links between economic instruments and changes in business behaviour is an imprecise science at the best of times (van den Bergh, 1996; Ekins, 1999; Baranzini, 2000). The first and possibly most irreducible problem is that of policy mix. As has already been noted, it is still relatively rare for economic instruments to be deployed without some accompanying regulation and, therefore, isolating the effects of fiscal incentives from those produced by other instruments can be extremely difficult (Goddard, 1995). The second problem is that of market complexity. Significant attempts have been made to incorporate less normative aspects of market preference into economic theory through such valuation techniques as hedonic, contingent and travel-cost pricing (Maddison *et al.*, 1996)[5]. However, although these techniques can simulate market situations with a reasonable degree of precision, business planners must take into account numerous market factors and the specific circumstances of their firm when evaluating responses to any form of policy. Thus, commercial considerations and the internal politics of the company, some of which may be outside the ambit of national regulation or subjective in character, form an important component of corporate environmental management and militate against a stylised relationship between economic instruments and polluter behaviour (Jones, 1999). Although such problems afflict all policy analysis, it is particularly pronounced for market-based mechanisms because they employ subtle and indirect price stimuli to promote greater environmental consciousness and action. The difficulties of defining the exact relationship between environmental taxes and industry behaviour are therefore very real and not easily circumvented.

In this study, the strength of the relationship between Green Dot and PRNs and corporate waste management strategies was examined by adapting a methodology developed by Labatt (1997a; 1997b), who correlated corporate waste reduction initiatives against various discretionary and mandatory policy measures established by Canadian provincial authorities. The approach adopted for the survey of UK and German packaging producers was to correlate corporate waste-management targets against companies' Green Dot or PRN costs. This was conducted for each target (reduction, re-use, collection and the purchase of packaging made from recyclate). Two consolidated measures were also devised to provide an overview of corporate engagement with the different forms of waste management examined in the survey. The first measure examined the number of waste management plans adopted by respondent businesses and was calculated by assigning each company a score of five points for each type of waste management plan and an additional point for each material included within each plan. The purpose of awarding additional points for each material was to differentiate companies that had adopted a re-use strategy for, say, just paper packaging waste from those with a re-use policy for all the main packaging materials covered by the Directive. Thus, if a business had adopted a waste reduction plan, it scored five points for the plan and a further six points if the plan included all packaging materials. The combined scores for waste reduction, re-use, collection and purchase of recycled packaging were then calculated to produce the first consolidated index.

The aim of the second aggregated index was to provide a synopsis of businesses' overall waste management targets, in order to produce an approximate picture of each firm's commitment to more active packaging waste management. This index was calculated by adding together the percentage targets set by respondents for each type of waste management examined in the survey. Companies that had developed targets for all areas of waste management measured therefore achieved a score equivalent to their combined percentage targets. To complete this element of the analysis, the individual and consolidated indices were then correlated against respondents' Green Dot or PRN charges. In order to assess the impact of company size and intensity of packaging consumption on the relationship between packaging waste charges and waste management activity, several permutations of the correlation were performed to measure environmental charges as a proportion of company turnover and as a percentage of total packaging consumption. In the event, neither factor significantly altered the results obtained.

It is recognised that this methodology is at best a crude means of quantifying the connection between economic instruments and business behaviour, as it assumes a linear and causal relationship between the two factors (Labatt, 1997a; 1997b). In reality, few economists would argue that the connection is this straightforward, especially where economic instruments are employed alongside other regulation or where multiple objectives are pursued. Equally, many businesses claim that increasing environmental costs rarely provides the chief impetus for change compared with standards-based legislation or supply-chain pressures (Beder, 1996; Hill, 1997; Watkins, 2000). These caveats clearly apply in this case, as the Directive seeks to promote various alternatives to the landfilling of packaging waste and has involved the introduction of several policy instruments. The tentative hypotheses, therefore, were that there would not be strong correlation between environmental charges and changes in waste management in either country but that, if iterations in producer levies are to impact upon polluter behaviour, the relationship should be stronger in Germany than the UK because Green Dot fees are substantially higher than PRN fees.

The results of the correlation analysis partly corroborated these hypotheses but less convincingly than anticipated (Table 5.5). Whilst there was a clear relationship between recycling charges and actions in Germany for some measures, there were some notable exceptions, including source reduction 1998-2001, re-use in 2001, and the combined waste management target index. Added to this, the overall differences between British and German companies were less than might have been expected given the differential in packaging waste charges and, moreover, statistically significant correlations were only observed for waste collection in either country. This suggests that economic instruments have not produced major changes in corporate waste management over and above those prescribed by national recovery and recycling targets. Notwithstanding the fact that the Green Dot and PRNs were designed primarily to generate investment funds rather than to alter industry behaviour, economic theory on environmental taxes nonetheless postulates that iterations in charges should affect the relative appeal of abatement compared with continued

Table 5.5   Waste management targets and environment charge costs

Waste Management Technique	Germany Spearman correlation	UK Spearman correlation
Collection 1998	0.215*	0.098*
Reduction 1998-2001	0.014	0.043
Re-use 1998	0.107	0.049
Re-use 2001	0.014	0.018
Buy Recycled 1998	0.101	0.013
Buy Recycled 2001	0.077	0.039
Index of all actions	0.101	0.037
Index of combined targets	0.023	0.033

* Significant at 95% confidence

pollution. The reasons for the weak relationship between environmental charges and producer behaviour in this instance therefore warrant further investigation and may provide useful insights into the wider use of economic instruments and other market-based mechanisms in environmental policy.

The most obvious explanation is that packaging waste charges are simply too low in either country to have a major impact on business behaviour beyond stimulating activity to meet legal requirements. In fact, this is consistent with economic theory on the marginal-cost attributes of incentive charges. Optimal abatement incentives are created with environmental taxes and charges where the marginal environmental-damage cost of further increases in pollution (in this instance, the production and consumption of packaging waste) is equal to the marginal benefit gained by the company from this activity. Clearly this point has not been reached in either the UK or Germany since the value gained by companies from selling packaged products exceeds the tax rate, which is only levied on packaging rather than the whole product (Pearce and Turner, 1992). Any incentive effect is therefore likely to be negligible. Moreover, inducing abatement incentives through such taxes could prove to be economically damaging and politically unfeasible, as it might require raising taxes to the point where inflationary pressures increased significantly.

In fact there are numerous examples where even punitive economic instruments have produced only marginal reductions in demand, especially in the case of price-inelastic commodities (Jacobs, 1991; Pearce *et al.*, 1995). For instance, the Conservative government in the UK introduced a fuel duty escalator in 1993 in addition to standard petrol duty as a means of raising tax revenue and discouraging road traffic on environmental grounds. The escalator was initially set at three per cent above annual inflation and was increased by the new Labour administration in 1997 to five per cent over the retail price index. However, by the time the Chancellor of the Exchequer scrapped the escalator in 2000, taxes comprised 81.5 per cent of total fuel prices, prompting blockades of oil refineries by freight haulage firms (British Broadcasting Corporation, 2000). Public opinion was generally supportive of these protests and reductions in fuel costs, such was British society's reliance on road transport, and debate on alternatives to road-based transport failed to capture the public imagination (The Guardian, 2000). In this example, the cost of a highly price-inelastic product became unacceptable before environmental taxes could induce a significant decrease in its use. In addition to highlighting that high environmental taxes do not always stimulate abatement, the case also suggests that the availability of alternatives (affordable and convenient public transport) is critical in determining the incentive impact of environmental taxes. Though the example of packaging waste is less extreme, logistics and marketing considerations do impose constraints on the abatement options available to firms operating national and international supply chains.

Conversely, international agreements to phase out ozone-depleting chlorofluorocarbons (CFCs) were concluded remarkably swiftly in the Montreal Protocol in 1987 because substitutes for CFCs were readily available and economically viable (Jacobs, 1991). Thus, although CFCs had previous been considered to be indispensable for refrigeration processes, technological progress as well as environmental consciousness dramatically increased the price elasticity of CFCs and facilitated their prohibition. In addition to reinforcing the importance of the availability of less environmentally damaging alternatives, the second example also illustrates that price elasticity can affect the efficacy of all methods of environmental regulation, not just economic instruments (Hahn, 1989; Halkos, 1996). Whilst the hypothecation of revenue from economic instruments can assist in providing investment in alternative technologies and capacity building, the examples underline that the introduction of economic instruments does not inevitably create an effective pollution reducing incentive (Carraro, 2001).

Notwithstanding the price elasticity of packaging, it is still logical to argue that environmental charges should provide individual firms with an additional incentive for small-scale reductions in packaging consumption. As noted earlier, many German companies have reduced their Green Dot and packaging costs by 'light-weighting' packaging or switching to materials that attract lower levies. Economic instruments can therefore be deployed progressively to stimulate industry's innovative and adaptive capacity. Numerous factors must be taken into account when assessing the environmental efficacy of the incentive structures in place in the UK and Germany. In both countries, recycling charges are calculated using complicated algorithms; PRN costs are determined by the type and weight of packaging, whereas Green Dot fees also take account of the volume and area of each packaging unit. However, there appears to be little commonality in the incentives created by recycling charges. The incentive for German companies, for example, is to sell bottled drinks in glass containers – which is considered to be less environmentally damaging by the Fraunhofer Institute because it is re-usable – rather than plastic, as glass attracts a charge of £47 per tonne compared with £890 for plastics, but is heavier by a factor of 11.67. The Green Dot charge is therefore equivalent to 2.7 pence for a 0.5 litre bottle for plastic and 1.6 pence for glass. Similar data for the UK reveal an average PRN cost for the same 0.5 litre bottle of 0.2 pence for plastics and 0.5 pence for glass[6]. The incentive for UK manufacturers is therefore to switch from glass to plastic, the opposite to that for German companies. This accounts for the decision by many UK drink manufacturers to increase production of plastic bottles whereas, in Germany, the re-use provisions of the Ordinance and Green Dot fees encourage greater use of glass bottles. Examples for other packaging materials (Figure 5.3) show an economic advantage for German manufacturers to switch from aluminium to steel containers but the opposite under the PRN scheme. Whilst there is no conclusive explanation for these discrepancies, the market-led PRN system might actually be undermining the environmental benefits of the Packaging Waste Directive if one accepts the life-cycle assessments of the Fraunhofer Institute. Moreover, it demonstrates that economic instruments have the potential to increase environmental damage if incentive structures are not based on sound ecological criteria.

A further explanation for the weak relationship between environmental charges and polluter behaviour is the fact that corporate actions are inevitably influenced by a host of exogenous commercial factors that can militate against effective incentive taxes (Jones, 1999). This point was further corroborated in interviews with UK and German packaging producers. For instance, one electronics manufacturer with representation in both countries stressed that as its annual compliance costs amounted to £30,000, any possible savings from re-evaluating the design and consumption of packaging were marginal to its overall business and did not justify major project expenditures. Other respondents acknowledged that PRN costs had increased the pressure for packaging re-design but felt they were overshadowed by other business considerations, such as the logistics and marketing benefits of packaging and

Figure 5.3 Unit recycling costs, Germany and the UK

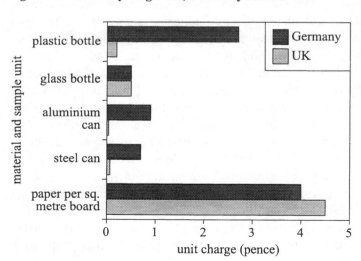

the standardisation of product branding throughout Europe. Simply put, businesses rarely make major operational commitments in response to relatively minor cost pressures. Other manufacturers claimed that as they routinely explored opportunities for packaging 'optimisation', neither legislation nor economic instruments had influenced their decisions. The British Retail Consortium's (BRC) submission to a review of the UK Regulations in 1998 echoes these sentiments (BRC, 1998: 3-6):

> *Many retailers have long been doing all they can to encourage recycling and minimise packaging use.* Examples include provision of banks on car parks; specification of recycled materials in packaging; packaging minimisation programmes; increased use of reusable packaging as demonstrated by closed-loop reusable schemes; and heavy investment in equipment for recovery of backdoor packaging waste. However, packaging is necessary in terms of product protection and health and safety considerations. *Reuse will occur if there is an economic benefit to it.* (emphasis added)

These responses underline Ekins' (1999) claim, noted in Chapter two, that it is vital for governments to understand the main objectives of new environmental taxes and charges when deciding on their structure and level. Although any increase in business costs may prompt some reduction in polluting activities and therefore all environmental charges have some incentive potential, the range of market and internal cost factors affecting responses to economic instruments can eclipse the potential for a noticeable incentive effect. Equally, if other forms of strategic behaviour promise greater financial rewards for individual firms – in essence, failures in policy design rather than the underlying principle of environmental taxes and charges – major shifts in production and consumption patterns are unlikely.

A final factor that has possibly militated against a strong relationship between economic instruments and polluter behaviour in this instance is the redistribution of environmental costs within the market. One of the intended functions of environmental taxes and charges is that they spread the costs of pollution through the economy in order to re-internalise externalities and prompt adjustments in production and consumption patterns (Smith, 1997). If, however, the dissemination of environmental costs to a large number of actors (suppliers, customers and consumers) merely succeeds in diluting the financial impact of the charge, the instrument may become a negligible consideration to all parties and fail to raise stakeholders' awareness of the environmental issue in question. This effect would be especially pronounced if, as with packaging waste, the environmental charge was only a minor component of the overall cost of the product in question. In the surveys, 58.9 per cent of German firms claimed that they had increased the price of packaged goods to include recycling charges compared with 32 per cent of UK companies. This suggests that as the level of environmental charges rise, the incentive to diffuse associated costs through the economy also increases. However, although 41.5 per cent of German respondents had informed customers of the reasons for price increases on packaged goods, reinforcing the message sent through price signals, only 21.1 per cent of British firms had followed suit. There is therefore greater evidence of cost diffusion serving a positive role in inculcating environmental awareness throughout the market in Germany than in the UK, again because of higher environmental charges. As with the overall effect of environmental charges, however, it is difficult to discern the effect of cost diffusion from that resulting from other informational devices, such as Germany's Recycling Awareness Days.

In summary, the implementation of the Packaging Waste Directive in the UK and Germany has revealed some important lessons on the use of economic instruments in environmental policy. Despite marked contrasts in the application of economic instruments in the two countries, neither PRNs nor the Green Dot has produced a strong incentive effect for changes in packaging waste management. The principal reason for this is that the charges required to achieve efficient funding for reprocessing and recycling are fundamentally lower than those needed to persuade industry to reduce, re-use, collect, and buy more recycled packaging (Smith, 1997). Economic instruments *may* have produced an overall reduction in packaging consumption in Germany but the main motivations for changes in waste management have been regulatory targets and the popularity of recycling amongst the German public. In the UK, economic instruments and legislation have produced little discernible reduction in packaging consumption. Packaging has become an integral component in the production, distribution and marketing of many consumer goods in both countries, making it a demand price-inelastic commodity. Other factors, such as anomalies in the relative pricing of different packaging materials and the diffusion of environmental costs through the market, have also militated against a straightforward relationship between economic instruments and corporate behaviour. In such instances, variations in economic instruments have created little additional incentive for businesses to engage with preventative waste management. For price-inelastic products and processes, therefore, the incentive effect of environmental taxes may be politically impracticable to pursue and hypothecation may offer greater potential for managing pollution through the establishment of closed-loop systems of taxation and mitigating expenditures. Equally, in many cases more traditional regulatory devices are likely to remain the principal instrument for establishing and realising the objectives of environmental policy, with economic instruments often serving a subsidiary role to overcome implementation difficulties, rather than operating as fully-fledged market-based mechanisms.

5.6　Alternative Policy Strategies

Notwithstanding their contrasting regulatory styles and differing environmental standards, there are strong similarities in the methods used by the British and German governments to implement the Packaging Waste Directive. Both used statutory instruments to transpose the Directive then deployed comparable NEPIs (negotiated agreements, economic instruments and hypothecation) to facilitate practical implementation. The British government's decision to incorporate a market-based tradable permit scheme into its implementation strategy was therefore the major methodological distinction between the two national policies. At the same time, the nature and performance of these policies and NEPIs have been strongly influenced by the regulatory styles of the two governments.

Practical implementation has not been an unqualified success in either country, however, as unforeseen complications have arisen, distorting incentives have been created, and the ability of the authorities to respond to these problems has been restricted by political preferences and the administrative structures created at the outset of implementation. There has thus been a degree of path dependency in both countries which has shaped the search for, and application of, policy solutions (Pressman and Wildavsky, 1984; Jones, 1997; Dimitrakopoulos and Richardson, 2001). Considering the difficulties experienced in implementing the Packaging Waste Directive using the regulatory and NEPI techniques employed in Britain and Germany, it is appropriate to consider whether alternative or refined policies might overcome some of these deficiencies. This is not to suggest that existing measures have failed in either country but rather that some objectives might have been achieved more easily if they had been approached from a different perspective.

During discussions on the implementation of packaging-waste policies, numerous options were considered by national governments and, indeed, the Dual System and producer responsibility only emerged after extended discussions between government, industry and other stakeholder groups (Haverland, 1999). The UK government was particularly keen to examine all policy angles that augured cost-effective compliance with the Directive. Further reflection also took place after the Regulations came into force as it became apparent that aspects of Britain's recycling system were under-performing. For example, the DETR briefly courted the idea of introducing separate recycling targets for industrial-commercial and domestic waste as a way of increasing the collection of post-consumer waste by reprocessors and compliance schemes (DETR, 1998b). Consumer taxes on packaging waste were also considered but both suggestions were felt either to be ineffective, economically damaging, or regressive for poorer sections of society (ACP, 1998; DETR, 1998c; 1999b).

Several of these and other alternative policies were assessed in the surveys to determine their feasibility and desirability from the perspective of industries affected by the current regulations (Table 5.6). In fact, even suggestions that might reduce the scale of producer responsibilities received little support from respondents in either country. First, nearly three quarters of respondents felt their sector had been treated even-handedly by government policies despite the general belief that current recovery targets are too high. Moreover, neither group felt that removing mandatory targets or charges would increase the efficacy of policy implementation, suggesting that although industry groups were wary of formal regulation, their main pre-condition for accepting producer responsibility was that it be applied in a way that deterred free riding (ACP, 1998). Similarly, respondents from both countries appeared to accept their obligations under the polluter pays principle and the reality that packaging-waste systems could not proceed without an adequate financing mechanism.

Similar antipathy was expressed towards the creation of recycled-content quotas for packaging. Though respondents acknowledged that this would

Table 5.6   Corporate opinions on alternative policy strategies

Proposition	Per cent of companies agreeing with proposition	
	UK	Germany
Industry treated unfairly by government policy	24.2	27.1
Volutary targets would be more effective than government regulation	25.9	31.3
Targets could be more effectively achieved without waste charges	25.1	14.0
Government should specify recycled-content quotas for packaging	38.7	36.0
The public should be directly taxed for packaging waste	42.0	25.5
Expenditure on public education should be increased	84.1	45.4

help to guarantee end markets for recyclate, mandatory quotas were seen as overly prescriptive and unenforceable by both groups. There was slightly greater support amongst UK respondents for public taxes on packaging waste to improve recycling (42 per cent), though only 25.5 per cent of Germany companies felt user taxes were preferable to producer charges. As noted earlier, the British government rejected public taxes for packaging waste on the grounds they could detrimentally impact on poorer households, as waste generation is not directly related to income (see also McQuaid and Murdoch, 1996; Ebreo *et al.*, 1999). The option enjoying greatest support was increased expenditure for public education on packaging waste and recycling. 84.1 per cent of British firms and 45.4 per cent of German companies felt that more resources in this area would improve recycling rates, though it should be remembered that a substantial proportion of Green Dot revenue is already diverted towards encouraging public participation in waste segregation.

In summary, neither group supported radical alterations to existing policies and, despite some misgivings about the economic or environmental efficacy of national packaging regulations, UK and German firms appear to be generally satisfied that current policies are non-discriminatory and have been implemented using appropriate policy instruments. In assessing the reasons for this general impression, it should be remembered that firms have invested heavily in developing recycling networks – German industry, for example, spent nearly seven billion *Deutschmarks* setting up the Dual System – and would be disinclined to see this outlay squandered. Uncertainty as to the direction alternative policies might take has also contributed towards industry's conservative attitude towards change. German industry has particular cause for concern, as the federal government's decision to re-introduce mandatory deposits for beverage containers from 2002 sent a clear message that if the existing arrangements failed to meet government objectives, adjustments to the Packaging Ordinance would further increase the burden on industry (Michaelis, 1995).

5.7   Conclusions

The emergence of NEPIs has heralded major changes in the way environmental policies are implemented in the EU. Although incentive taxes, cost-covering charges and voluntary agreements have long been part of the regulator's armoury in other areas of policy, their deployment in the environmental sphere is still comparatively recent and, consequently, policy-makers' experience in designing environmentally, economically and socially effective NEPIs is still limited (Dovers *et al.*, 2001). The translation of NEPIs from theory into practice has been a complex process and it is not surprising that mistakes have been made, especially where competing political priorities must be reconciled or the behaviour of individuals and markets is difficult to predict. The Packaging Waste Directive has proved to be a particularly interesting example of the ways in which such conflicts are negotiated because of the range of NEPIs deployed by member-state governments.

A number of broader lessons for the use of environmental charges and taxes have emerged from the case study. Although recycling charges in the UK and Germany were not conceived primarily as incentive taxes, the effects of environmental charges on corporate waste management were still less than might have been anticipated. This indicates that changes in industry behaviour have been prompted by the overall regulatory regime in each country rather than the specific effects of economic instruments, whose impact has been limited to their original cost-covering function. The lack of a significant incentive effect from the Green Dot and PRN environmental charges is explained mainly by the fact that the charge rates needed to raise revenue to meet the Directive's statutory targets proved to be lower than those required to create a discernible pollution-reducing incentive. Additionally, because packaging is a relatively price-inelastic commodity, only small reductions in demand have resulted from quite significant iterations in the scale of charges. These conclusions are generally consistent with economic theory, which notes that there is no logical reason why the optimal rates for incentive taxes and cost-covering charges should naturally coincide because of the different functions they seek to fulfil (Ekins, 1999).

Whilst the issue of price elasticity is well understood, another factor influencing the incentive effect of environmental taxes and charges is the threshold at which firms are prepared to invest in pollution abatement. Where such transaction costs are high, taxes and charges may need to rise significantly before discernible abatement activity occurs. Once this threshold is reached and new technologies or processes are developed, more rapid rates of abatement may occur. Finally, the apportionment of environmental costs through the market must be considered. Economic theory suggests that as environmental taxes and charges increase, polluters will seek to push a proportion of their compliance costs on to suppliers and customers, thereby helping to inculcate greater environmental awareness across the market. However, in some instances, the act of cost diffusion may dilute the message transmitted by price signals to the point where environmental costs become a negligible consideration for all elements of the pollution chain. In such cases, abatement may not increase with the application of an economic instrument.

For a variety of reasons, therefore, the incentive potential of environmental taxes and charges may be restricted and where one or more of the above conditions prevails, economic instruments may need to be deployed in conjunction with other policy tools in order to achieve desired objectives. Nonetheless, economic instruments can still be used in such circumstances to stimulate innovation and increase the price elasticity of commodities (Carraro, 2001). It is therefore important for decision-makers to appreciate fully the nature of the product or process being regulated before deciding on policy objectives and type of economic instrument to deploy. In many areas of environmental policy the hypothecation of proceeds for economic instruments can be an important component of strategies seeking to regulate price-inelastic commodities or processes. A number of schemes with similar objectives have already been developed in the UK. One such example is the Landfill Tax Credit Scheme, where levies from waste disposal charges are used for research on the amelioration of environmental impacts arising from landfill sites (Porter, 1998). The UK government has also earmarked £100 million a year from its Climate Change Levy for investment into renewable energy, with the objective of generating 10 per cent of power from renewable sources by 2010 (Terence O'Rourke plc/ETSU, 2001).

However, the Climate Change Levy represents a fundamental departure from many hypothecated taxes and charges, in that its investments aim to promote preventative action, namely the reduction of dependence on fossil fuels. By contrast, UK and German recycling charges are essentially remedial in that packaging waste must be produced and recovered before revenue is generated for investment in recycling infrastructure. By incorporating long-term prevention into the design of hypothecation schemes, fiscal measures can be used to regulate the demand for price-inelastic products. One ideological drawback of this approach is that it requires increased government intervention in the management of environmental taxes and charges at a time when the general preference is for more light-handed regulation. Conversely, as the previous chapter argued, inadequate control over hypothecation can expose mismatches between private incentives for firms and the objectives of environmental policy. Some government involvement in rule setting for market-led hypothecation is therefore necessary if such schemes are to attain both economic and environmental efficiency, though excessive intervention can lead to economic inefficiencies and impair business competitiveness.

Ten years on from the first of their *Blueprint* series, Pearce and Barbier (2000) reflect on the changes that have occurred in environmental policy since its publication. They conclude that environmental economics has become a common language for scientist and policy-makers alike and that its opponents are fewer as many have realised some virtue in the economic approach to environmental policy. However, they concede that putting the economic message into practice has proved to be more difficult than anticipated, as it has necessitated changing unsustainable institutions that have been built up over many years. As an addendum, they warn that the focus should now be on the task of applying environmental economics rather than on reconstituting the problem. The results of this study generally support these conclusions but also highlight some of the practical challenges faced when putting environmental economics into practice. In particular, they have presaged the dangers of over-stylising the relationship between polluter costs and behaviour in complex market situations and also of expecting economic instruments to achieve multiple functions. As a result, policy instruments must be carefully selected and their objectives clearly defined in advance of policy implementation. The hypothecation of cost-covering charges can help to overcome some of the difficulties caused by price-inelastic commodities but such strategies should also incorporate measures to combat the production of pollution rather than merely ameliorating its effects if policies are to maintain their focus on environmental sustainability.

In conclusion, the review of the implementation of the Packaging Waste Directive has drawn some interesting conclusions on the use of economic instruments, tradable permits and voluntary agreements for promoting more sustainable waste management. The results of the analysis generally support the contention that NEPIs have the potential to make a major contribution towards the implementation of EU environmental policies by reducing compliance costs and integrating environmental considerations into the decisions of the market. At the same time, it has highlighted critical factors determining the success of their deployment. The next chapter now re-examines the utilisation of NEPIs in the EU and, in particular, their ability to promote greater harmonisation of environmental policies in the member states and the relationship between NEPIs and the Single Market.

Notes

[1] For example, a retailer consuming 52 tonnes of packaging waste annually is required by the Regulations to recover 14 tonnes (52 tonnes x 56 per cent recovery target x 48 per cent sector obligation) but by reducing consumption to 49 tonnes, the company would no longer be required to recover or recycle any packaging waste.

[2] Moves are currently afoot in Britain and Germany to implement European Committee for Standardisation (CEN) standards on the design of packaging to satisfy the Essential Requirements of the Packaging Waste Directive. As yet, only two of these standards have been published by the Commission, covering the recyclability and combustibility of packaging but further standards are foreseen to encourage optimisation of the quantity of materials used to package goods (Raymond Communications, 2001).

[3] GKN operates a nation-wide system for the rental and re-use of wooden pallets. Participating companies are required to report each month on the number of GKN pallets they have in stock along with the number of receipts and despatches. If a company's stock of pallets is significantly lower than the calculated balance taking into account previous stocks, receipts and deliveries, it may be required to reimburse the hirer for the value of the pallets.

[4] The author was personally involved in developing a tote-bin system for a retailer seeking to reduce its consumption of single-trip cardboard packaging. 45,000 returnable tote bins were purchased and delivered to suppliers free of charge for the carriage of small items. Other retailers, particularly supermarket chains, have adopted tote bins for fresh produce and bakery goods.

[5] Pearce *et al.* (1989) review various methods of environmental valuation, including direct, indirect, hedonic, contingent, and travel-cost pricing. Direct valuation uses surrogate markets, where buying and selling processes are used to attribute values to environmental resources. Indirect valuation is based on a 'dose-response' technique, where estimates of reduced pollution or resource consumption are calculated and adjusted to achieve an optimum balance between cost and benefit. Hedonic approaches *estimate* how much of a property differential is due to a particular environmental variable, then *infer* how much people are willing to pay for an improvement and the social value of that improvement. Contingent valuation *directly asks* people what they are prepared to pay for an environmental benefit or accept as compensation for its loss. Travel-cost techniques assess how long and often people are prepared to pay to travel to an environmental amenity (and their duration of stay), based on the opportunity cost of the revenue gained and utility lost from going to work.

[6] The average weight of a 0.5 litre plastic bottle of carbonated soft drinks is 30 grammes. This equates to 33333 bottles per tonne. The weight of a comparable glass bottle is 350 grammes, with 2857 bottles per tonne (DETR, 1997b).

Chapter 6

NEPIs and EU Environmental Policy

6.1   Introduction

Although the limitations of traditional command-and-control approaches to environmental policy have provided the principal motivation for the recent upsurge in the use of NEPIs in the member states, their rapid dissemination is also symptomatic of broader political trends. First, it demonstrates the inroads into policy networks made by epistemic communities whose research espouses the merits of more flexible and efficient environmental policy instruments (Jordan *et al.*, forthcoming). Whilst there is little evidence that policy makers are anxious to relinquish control of key decisions, they have increasingly sought the advice of epistemic communities, particularly for solutions to environmental problems with a significant international dimension or major economic implications (Haas, 1992). Where such communities have been granted privileged access to decision-takers, this has, on occasions, enabled them 'to gain a stranglehold' on the way national policy is framed and expedited (Walker, 2001: 278). Jasanoff (1996) argues that this authoritative position may not always be entirely justified if the views expressed by leading epistemic communities become thought of as orthodoxy when, in reality, all scientific methods and prescriptions are based on subjective social constructions and beliefs. Notwithstanding this, NEPIs have also found favour because they reflect a more general political reaction in EU states against the over-regulation of government activity. Consequently, scientific arguments and political factors have both encouraged a greater receptivity amongst a sizeable proportion of Europe's political elite towards new ideas such as market-based instruments and voluntary agreements.

It would be erroneous to claim that standards-based legislation has been even remotely eclipsed as a *modus operandi* of EU environmental policy, however. Despite a slowdown in new environmental legislation, the directive remains the customary form of EU environmental policy and, even where member states have deployed NEPIs, this has generally been in conjunction with command-and-control regulation. Such 'mixed' strategies, where regulators specify standards but not how they must be achieved (Russell and Powell, 1996 in Pearce and Barbier, 2000: 197), has allowed national governments to experiment with new policy instruments whilst retaining the 'guaranteed' outcomes that traditional regulation is generally thought to provide[1]. In many instances, this has led to NEPIs being used as subsidiary mechanisms to address technical issues that cannot be adequately dealt with through legislation rather than as true market-based instruments. This has sometimes led to NEPIs that bear little resemblance in either style or substance to policy instruments of the same name in the economics literature. Where mechanisms of this nature have 'under-performed' against environmental policy objectives, their shortfalls are often attributed to regulatory failures rather than deficiencies in the theories underpinning NEPIs (Goddard, 1995).

Having reviewed these issues in the context of the Packaging Waste Directive, the purpose of this chapter is to examine the politicisation of NEPIs within the EU. The first section assesses the influence of EU membership on the design and operation of NEPIs, paying particular attention to the internal market and the need to ensure that nationally implemented NEPIs do not restrict the free movement of goods and services between member states. As national NEPIs become more popular, the Commission may need to scrutinise the competitive effects of such policies more closely in order to prevent fraying of the internal market. Member states' *a priori* awareness of the need to abide by EU trade requirements may also affect the design of NEPIs. At the same time, trade imperatives must be balanced against treaty limitations on EU intervention in the affairs of its member states. The second section takes the opposite perspective, assessing the influence of NEPIs on the EU environmental programme and its ambition to promote sustainable development. Controlled diversity has been a constant feature of EU environmental policy in recent years, particularly during practical implementation, but a key question is whether NEPIs are encouraging greater convergence or divergence of national policies. By the same token, the mere event of policy convergence does not guarantee that national environmental policies will become more sustainable even though the balance of evidence does suggest that NEPIs are a more environmentally and economically effective means of pursuing this objective than standards-based regulation (Pearce and Barbier, 2000).

6.2 The Influence of the EU on the Deployment of NEPIs

It was noted earlier that NEPIs are rarely deployed in their textbook form but are often modified by national governments to cater for practical exigencies and to conform to existing political and administrative structures. This pragmatic approach has the obvious appeal of increasing the chances of 'successful' implementation in terms of achieving stated policy objectives. However, it may lead to outcomes that fall short of the 'optimal' balance between environmental protection, economic development and social welfare envisaged by economic modellers. Pressman and Wildavsky (1984) consider this focus on practicality over theoretical consistency to be pivotal, arguing that successful implementation requires linkages to be developed across a large array of actors whose co-operation is needed to transform paper policies into tangible outcomes. Although obtaining the participation of a single actor at any given 'decision point' may be reasonably straightforward, the chances of maintaining favourable coalitions diminish markedly as the number of decision points and interests increase (Jordan, 1999). Pressman and Wildavsky (1984: 6) further warn politicians not to promise what they cannot deliver as to do so only leads to 'disillusionment and frustration' with the policy process.

The implementation of environmental policy is particularly intricate because it seeks to address complex and systemic problems. Whilst the majority of economic theory underpinning market-based mechanisms assumes perfect competition and rational market behaviour, governments are often called upon to devise NEPIs to regulate monopolistic or imperfectly competitive markets where 'standard' economic assumptions may be heavily distorted or simply not apply (Hahn, 1989; Nannerup, 1998; Carlsson, 2000). The information available to regulators for predicting market behaviour is also often less than comprehensive and, thus, decisions based on literal interpretations of economic theory are likely to produce unintended, and sometimes unfavourable, consequences (Lévêque, 1995). The need to adapt NEPIs is therefore a political fact of life and not restricted to the EU.

Nevertheless, when member states wish to introduce NEPIs they need to ensure these instruments do not unduly disrupt the free movement of goods and services within the Single Market. This obligation is enshrined in Article 28 of the EC Treaty:

Quantitative restrictions on imports and *all measures having equivalent effect* shall be prohibited between Member States (emphasis added).

Whilst quantitative restrictions are clearly incompatible with the notion of a common market, the words 'measures having equivalent effect' have been given a wide interpretation in order to remove intra-Community trade barriers (Wiers, 2002). If NEPIs applied at a national level restrict imports from other member states in either an open or hidden way, this increases the likelihood of conflicts between the effective and cost-efficient implementation of environmental policies and the Treaty's trade objectives.

However, since the official incorporation of environmental protection into the objectives of the Treaty, national policies that restrict free trade may be deemed permissible under Article 30 provided three conditions are met. First, the measure must not constitute a means of arbitrary discrimination or a disguised restriction on trade between member states. Second, it must be demonstrated that the policy is indispensable to protect the environment and/or public health. Finally, the policy must be proportional to the scale of the environmental problem being addressed (Heyvaert, 2001). The tension between environmental and trade policy is again not unique to the EU, as similar protocols exist to defend trade liberalisation in the World Trade Organisation (WTO) Agreements. It is nonetheless important to recognise that there is no institutional equivalent to supranational bodies like the Commission, European Parliament and ECJ in the WTO, nor does the WTO Secretariat function as the guardian of treaties in the way the Commission operates in the EU (Wiers, 2002). Equally, the depth of economic integration expressed in the EU treaties far exceeds that contained in the WTO Agreements. The rules governing trade and environmental policy in the EU *are* therefore unique in their scope and complexity (Jordan, 2002).

Whilst compliance with EU trade rules is the single most important issue national governments have to contend with when designing NEPIs, other areas of EU policy, such as social policy, agriculture and monetary union, also have a major bearing on the application of environmental policy. Equally, the term NEPIs encompasses a diverse assortment of policy instruments not restricted to environmental taxes, tradable permits and voluntary environmental agreements. An exhaustive review of the effects of the EU on NEPIs is therefore beyond the scope of this chapter. Rather, the aim is to provide a preliminary exploration of the ways in which the internal market may impact upon the design and deployment of NEPIs in the member states, based on existing knowledge of their characteristics and EU trade rules. The effects of non-trade requirements on NEPIs is nonetheless an issue which clearly warrants further research to understand better the relationship between the forces driving EU political and economic integration and those promoting sustainable environmental policy.

6.2.1 Environmental Taxes and Charges

Trade difficulties may arise for environmental taxes and charges where different tax rates are administered by member states for the same product or process, altering their relative cost over and above differentials created by normal factors of production. This would not automatically trigger infringement proceedings, however, as national governments retain the right, under Article 93, to determine their fiscal policies provided they are non-discriminatory, necessary and proportional. Moreover, a member state is unlikely to gain a competitive advantage in the Single Market from the introduction of new or higher environmental taxes[2]. The Commission would therefore have little incentive to intervene unless quantitative restrictions were also imposed on imports as part of the policy (Debelke and Bergman, 1998). The situation would be more complicated if other governments subsequently introduced environmental taxes covering the same issue at a significantly lower rate, in order to be seen to be protecting the environment but, at the same time, seeking to fulfil other, principally economic, agendas. Whilst the Commission might seek to overturn such taxes as pre-meditated protectionism, it would need to prove that the manoeuvre was deliberate and that it was disproportionate when social and economic conditions in the 'offending' state(s) are taken into account. Furthermore, the way in which revenue from environmental taxes is used must also be considered. Many levies are designed to be 'revenue neutral', that is, tax proceeds are recycled back to industry or used for specific environmental purposes, such as support for environmental investments. Although revenue recycling is consistent with the principles of ecological tax reform and an important component of environmental capacity building, both of which are supported by the EAPs, it may be construed as state aid to domestic industry, which, in principle, is forbidden by the Treaty (Debelke and Bergman, 1998). Permission for national hypothecation schemes must therefore be sought from the Commission and certain requirements have to be met[3].

Overall, therefore, the various permutations of environmental taxes and charges that can exist make their assessment against the internal market extremely difficult and it is perhaps understandable that the Commission has not directly confronted this issue. If the Commission did feel it necessary to compare national environmental taxes and charges between states rather than considering each tax individually, this could lead to protracted disputes over legal interpretation. Whilst Debelke and Bergman (1998: 249) report no major ECJ cases on environmental taxes or charges, the principles of non-discrimination and proportionality are sufficiently ambiguous that more formal systems of control may be needed to mitigate the international competitive effects of national environmental taxes. One possible means of achieving this, which is generally consistent with the EU's confederal structure, is the implementation of border tax adjustments (BTAs) (Ekins; 1999; Ekins and Speck, 2000). BTAs involve the reimbursement of taxes paid by domestic producers for exported products and the imposition of equivalent taxes on imported goods, so as to neutralise any competitive distortions[4]. There are a number of potential problems with BTAs, however. Their environmental benefits may be questionable if, for example, the remission of taxes for exports actually causes an increase in the production of an environmentally harmful commodity (Ekins and Speck, 2000). If, however, BTAs lead to a reduction in overall consumption, they may yield environmental as well as competitive benefits.

The extent to which BTAs are compatible with international trade rules is also not certain, as BTAs can themselves be used for economic protectionism (Ekins, 1999). Whilst product BTAs – such as charges on packaging waste – are considered generally acceptable by the WTO provided taxes are non-discriminatory within each product category, it is less receptive to BTAs for imputed input taxes on energy because of the difficulties in obtaining comparable valuations within product groups (OECD, 1996). A final objection to BTAs is that they run contra to the spirit of the internal market and are more appropriate for the global trading system. Nonetheless, such devices are worthy of consideration to help reduce the competitive implications of nationally applied environmental taxes (Hahn, 2000).

Other initiatives suggested for reducing the competitive effects of environmental taxes include the development of common rules for recycling revenue to ensure taxes remain fiscally neutral (Baranzini et al., 2000). However, any such scheme is likely to produce winners and losers, such that fiscal neutrality may be possible at a national level but not for individual firms or sectors. Another mitigating measure might be uniform rules for tax exemptions and rebates ceilings, comparable those already applied in the UK under the Packaging Regulations for smaller businesses (see Chapter three). However, the adoption of either measure on a systematic basis implies a level of EU intervention in national fiscal affairs that is unlikely to be considered acceptable by many member states.

6.2.2 *Tradable Permits*

Tradable pollution permits raise similar issues for the internal market as environmental taxes and charges, since both apply fiscal incentives for polluters to engage in more environmentally responsible behaviour. Pollution permits applied on a national basis can therefore alter trading conditions between domestic firms that have joined a scheme and foreign competitors who may be excluded from the initial permit issue. For instance, companies may use the proceeds from permit sales to make their products less expensive but, more importantly, those purchasing permits would obtain the *de facto* right to pollute at a higher than average rate, thereby evading abatement costs in the short term, although they would still incur the cost of acquiring additional permits. This might occur where an EU pollution standard is established but, as a result of a national permit scheme, the standard is flexibly apportioned between polluting companies in that state (Joyce, 2002). The competitive advantage created by permit trading would diminish in the longer term if the scheme's administrator periodically withdrew permits from circulation in order to stimulate advances in abatement technology and techniques.

However, several features distinguish tradable permits from other forms of MBM and make their trade implications worthy of separate consideration. First, tradable permits do not necessarily require an initial financial outlay from polluting businesses. Under the US sulphur dioxide trading scheme, which began in 1990, permits were issued free of charge to polluting industries to stimulate market-based environmental improvements without the imposition of additional transaction costs on industry (Joyce, 2002). Even where permits are auctioned, the regulatory charge is normally a one-off payment or a renewable lease. Thereafter any costs to polluters stem from trading between firms that reduce emissions and those seeking to pollute at higher levels. Although both groups may incur ongoing costs, these decisions are guided by independent appraisals of the costs and benefits of abatement *vis-à-vis* continued pollution rather than by regulatory mandates[5]. By contrast, environmental taxes and charges apply a continuous cost in order to encourage reductions in pollution. A third source of additional cost, the withdrawal of permits from the system, is a regulatory pressure though, again, market forces determine how firms react. Whether this constitutes a significant departure from environmental taxes is not entirely clear, however, as firms responding to incentive taxes may reduce pollution sufficiently that they also gain an overall financial benefit. This would occur, for instance, where companies reduce energy consumption in order to limit a carbon tax liability. Despite this similarity, the incentives used to encourage abatement vary appreciably between the two instruments in the degree to which they employ financial 'sticks' or 'carrots' to encourage greater environmental stewardship (Jacobs, 1991).

Another distinctive feature of tradable permits is the need for regional organisation for some forms of pollution. In the case of sulphur dioxide emissions, if unrestricted trading is permitted over a wide geographical area, this can result in localised concentrations of permit holders, causing pollution 'hot spots' and social and health inequalities. The customary means of avoiding this is the division of permit quotas into sub-regions (Atkinson and Tietenberg, 1991). Environmental taxes do not tend to create regional concentrations in pollution, since they are applied on a sectoral rather than a geographical basis though, as noted earlier, low environmental taxes in one member state might encourage polluting industries to locate there. Whether or not tradable permits create regional disparities also depends on the type of pollutant being regulated. The majority of existing schemes have focused on air pollutants, such as sulphur dioxide, volatile organic compounds, carbon monoxide, nitrogen oxides and lead reduction, all of which produce severe localised impacts (Golub, 1998). The application of tradable permits to greenhouse gas emissions would have fewer spatial implications, since greenhouse gases impact upon the global climate system irrespective of where they are emitted but are not all noxious pollutants. Issues of regional organisation may nonetheless apply to waste management schemes if they lead to increases in waste disposal in a few areas. This might conceivably occur with the trading scheme proposed by the UK government as part of its plans to implement the EU Landfill Directive (1999/31/EC). This scheme aims to reduce the amount of biodegradable waste going to landfill sites by issuing each local authority with disposal permits, then withdrawing permits at intervals in line with EU targets[6]. The government hopes that the scheme will encourage authorities which can economically compost a greater proportion of their biodegradable waste to sell excess permits to those experiencing difficulties meeting targets, thus reducing overall national compliance costs (DETR, 2000a). However, the risk of increased waste disposal in some areas is, as yet, hypothetical, and there is minimal evidence of regional impacts arising from the PRN scheme.

Whilst tradable permits have the potential to distort trade between member states, decisions on whether or not to reduce pollution are based on each firm's evaluation of the cost and benefits of abatement and, therefore, the Commission would be unlikely to question the competitive effects of such schemes. However, the eligibility rules attached to some trading schemes might be construed as more clearly discriminatory. Most schemes restrict the right to hold pollution permits – usually to certain industry sectors and their agents – in order to prevent outside parties speculating on their value. Indeed, rumours of profiteering seriously undermined the credibility of the PRN system in its early years because the UK government did not initially impose a mechanism to restrict ownership of PRNs to industries affected by the Packaging Regulations (see Chapter three). Membership rules are therefore necessary to ensure the effective operation of permit markets but, nonetheless, schemes applied on a national basis can deny firms from other member states the opportunity of competing on an equal basis with domestic industries.

This tension between the internal market and the development of an effective trading system is difficult to resolve through general principles and may be a contributory factor in the slow uptake of tradable permits in the EU compared with the USA (Golub, 1998). However, at the time of writing, the Commission has not objected to a carbon-trading scheme introduced by the UK in April 2002. The initial auction of permits was held in March 2002 for 34 companies and more are expected to join, though the scheme specifically excludes companies from the utilities sector. In order to increase the incentive for reductions in carbon emissions, the government has made available an incentive fund of £215 million for the next five years and the scheme is expected to reduce carbon dioxide emissions by four million tonnes over four years (DETR, 2000b; UK Research Office, 2002). A number of features of the UK scheme render it unlikely to be the subject of ECJ proceedings. First, the scheme is limited in scope and, therefore, the size of the trade constraint is certainly not disproportionate to the scale of the environmental problem being addressed. Second, by earmarking funds to stimulate innovation, the scheme seeks to protect the environment, an essential objective of the EU Treaty. Finally, climate change is an issue which is not comprehensively regulated by EU legislation and, therefore, applying the precedent set by the Danish Bottles Case, member states have the right to adopt national measures if they are deemed essential to environmental protection (Wiers, 2002). This situation may change in the near future, as the Commission is currently evaluating options for an EU trading scheme to reduce the costs of implementing the Kyoto Protocol. Capros and Mantzos (2000) suggest that, if each member state implements its target under the EU's Burden Sharing Agreement individually, the annual cost will be Euro 9.0 billion but that an open trading system across all member states and sectors would reduce compliance costs by Euro 3.0 billion per annum by 2010. If the Commission's proposals for EU carbon trading come to fruition then the British government may be pressured to modify its scheme.

Whilst an EU carbon trading scheme has strong environmental and economic justifications and would eliminate the possibility of indirect protectionism, EU trading systems for other pollutants would need to take into account the issue of regionalisation. One option would be to develop framework directives to set overall objectives and sanction the use of tradable permit schemes for particular pollutants. Day-to-day management of these schemes – including the auctioning of permits, regional organisation and monitoring – would then be delegated to the member states. This approach would have the critical advantage of encouraging more member states to take up tradable permit schemes, thereby reducing the overall economic impact of environmental policy and the competitive effects created by isolated national schemes. Even then tradable permits may not be appropriate for all member states if the industry being regulated is a monopoly or oligopoly. The decision whether or not to adopt tradable permits to meet a particular objective would therefore need to remain at the discretion of each member state. More formal systems of control, such as BTAs, would be extremely problematic to enforce, however, because of the difficulties in quantifying the savings accruing from permit trading compared with the costs of other implementation methods.

6.2.3 *Voluntary Environmental Agreements*

Voluntary environmental agreements are generally considered to be compatible with the internal market notwithstanding the fact that companies participating in an agreement might incur additional implementation costs, since it can normally be presumed that they have acquiesced to environmental standards which do not impose a greater competitive disadvantage than would have occurred with more formal regulation. This was certainly the case with the producer responsibility agreements used in the UK and Germany to implement the Packaging Waste Directive (see Chapter three). Other member states are also unlikely to complain if their industries stand to gain a competitive advantage from another country's voluntary environmental agreement. This view is supported by case experience, which reveals no instances where member states have made compliance with standards set by voluntary agreements a pre-condition for imports from other countries (Debelke and Bergman, 1998)[7]. A final reassurance for the Commission may lie in the fact that many, though not all, negotiated agreements do not represent a radical change in existing regulatory frameworks (Karamanos, 2001). Instead, they are frequently used as auxiliary instruments to implement EU environmental law and are consequently an extension of the current regulatory arena rather than a dramatic departure from it (Lober, 1997). The agreement on the formation of the DSD in Germany nonetheless demonstrates that corporate voluntarism is not always a guiding principle, as the DSD was very much a defensive strategy on the part of German industry in the face of intense regulatory pressure. Moreover, the diversity of voluntary agreements in terms of their legally binding obligations (or lack thereof) militates against simple adjudication on their competitive effects.

In addition to the normal conditions used by the Commission and ECJ to assess the competitive effects of voluntary environmental agreements, the Court may examine the extent to which the agreement promotes the systematic replacement of an environmentally damaging product by a more acceptable alternative. This has become known in EU case law as the substitution principle. In the *Toolex Alpha* case in 1998 (C-473/98), the Court upheld a

Swedish prohibition of the industrial use of trichlotoethylene (TCE) on the grounds that there was scientific evidence that the substance was carcinogenic and because less harmful alternatives existed that could be commercially developed. The Court therefore felt that the TCE ban was justified even though it failed to meet normal proportionality criteria because it promoted the systematic removal of a substance considered to be damaging to public health.

Following this judgement, the ECJ has invoked the substitution principle with increasing regularity to settle disputes as to when, and to what extent, restrictions on free trade should be granted on the grounds of environmental protection (Heyvaert, 2001). This has been deemed necessary because justifications for trade restrictions based on their indispensability and proportionality have proven extremely ambiguous to interpret in law. Heyvaert considers the *Toolex Alpha* to be particularly significant, arguing that the Court used the substitution principle effectively to lower the hurdles for compatibility with the requirements of necessity and proportionality for national policies that restrict free trade in order to promote environmental protection.

It would be less straightforward to apply the substitutability principle to environmental taxes and tradable permits, however, as the Commission insists that restrictive policies should usually contain targets indicating a definite shift towards more benign production and consumption patterns. By contrast, taxes and permits merely encourage substitution through administered adjustments to market prices and, thus, do not necessarily provide sufficient evidence of a systematic substitution process unless a sizeable amount of revenue is being earmarked for the development of less damaging alternatives. Two cases where this principle has been applied in the UK are the Climate Change Levy, where approximately £100 million per year is being invested in promoting renewable energy, and the Landfill Tax Credit Scheme, which finances research to help mitigate the environmental impacts of landfill sites. The Green Dot and PRN schemes have also helped to deliver noteworthy increases in recycling capacity in the UK and Germany, though it is questionable whether the PRN system's reliance on market-led investments represents a systematic substitution process.

Developing formal mechanisms to remove trade restrictions caused by voluntary agreements equivalent to those suggested for environmental taxes would be problematic because agreements can incorporate various kinds of objectives, incentives and procedures. Uniform rules might even prove counterproductive if standardisation leads to preconditions that are inappropriate for some member states or regions. The application of BTAs, for instance, would raise similar problems as those for input taxes, because of the difficulties in developing common criteria to compare the costs incurred by industries as a result of national voluntary agreements. The Commission has therefore tended to rely on the proportionality principle to assess the competitive effects of voluntary agreements, with the further proviso that consumers should achieve a substantial share of the benefits arising from any environmental agreement (CEC, 1996b). Heyvaert (2001) nonetheless suggests that the ECJ may opt to base more rulings on the substitution principle because of the ambiguities of interpreting the proportionality principle. Whether this will be applied directly to disputes on voluntary environmental agreements remains to be seen but, in the absence of firmer rules, it seems to offer useful guidance. At the same time, the substitution principle is not a panacea for resolving trade restrictions caused by voluntary agreements, as the ECJ needs to be able to decide, in each case, whether the action constitutes a systematic substitution process and whether, in reality, it reduces or increases environmental degradation. Ultimately, any ruling based on environmental principles tends to be a precarious exercise (Heyvaert, 2001).

In summary, the balance of evidence from this preliminary review suggests that the competitive impacts of EU membership on the design and deployment of NEPIs have not, thus far, been extensive. There may even be potential synergies between NEPIs and the internal market. If, as economists claim, NEPIs provide equal or superior environmental protection at lower cost than command-and-control, their spread could reduce the economic impact of national environmental policies and, hence, their competitive effects. Whilst the requirement to ensure that NEPIs remain compatible with the internal market has the potential to cause frictions within EU environmental policy, the trade restrictions caused by most NEPIs are not especially major and, therefore, should not restrict their adoption by national governments. This is not to deny that NEPIs have been deeply politicised by EU integration but as the case of the Packaging Waste Directive demonstrates, the driving force behind their modification has been domestic political pressures and preferences.

In a similar vein, the difficulties in developing more than general principles to regulate the competitive effects of NEPIs suggests that adjudication will continue to take place predominantly on a case-by-case basis. The key exception to this is environmental taxes and charges, where more formal BTAs might be introduced to control trade distortions. Even this option is restricted to product charges, whereas many governments have sought to push environmental costs further up the pollution chain, using input taxes to improve the integration of environmental values into market prices. Moreover, despite a proliferation of environmental taxes in recent years, the Commission has shown little appetite for BTAs because of restraints on its ability to interfere in national fiscal policies. The undertone of disapproval emanating from the WTO has also dampened the Commission's enthusiasm for BTAs. Rather, it has tended to emphasise the benefits of environmental taxes at a supranational level, albeit with little apparent success. Having reviewed the competitive effects of NEPIs, the discussion now examines their broader implications for the nature and direction of the EU environmental programme.

6.3   NEPIs and the EU Environmental Programme

It is now generally accepted that the emergence of NEPIs has sparked a major transition in EU environmental policy (Golub, 1998). Whilst the search for flexible and efficient policy instruments at a supranational level has been hampered by concerns about over-federalisation of environmental policy, innovation at state level has been far reaching[8]. Reviews of NEPI performance in practice have also been generally positive, raising hopes that they will make a major contribution towards improving the implementation of EU environmental policies (see, for example, Ekins, 1999; Pearce and Barbier, 2000; Folmer *et al.*, 2001). It may nevertheless be several years before the full impact of NEPIs on the EU environmental programme is known, particularly where they rely on price signals which take time to filter through to market and consumer behaviour. Even so, it is important that the ramifications of NEPIs for European integration are thoroughly explored at the earliest opportunity. First, it is important to understand whether NEPIs are likely to encourage greater convergence of national environmental policies as member states adopt policies that have proven successful in neighbouring countries or whether the flexible nature of NEPIs will actually result in heightened diversity. Although the notion of flexibility has become accepted as an inevitable and sometimes beneficial aspect of EU environmental policy, excessive divergence may cause serious trade disputes and undermine progress towards the objectives of the EAPs. Second, further information is needed on the environmental consequences of NEPIs, in particular to determine whether a reduced emphasis on standards-setting and prohibitions might erode environmental protection across the EU or encourage a more comprehensive integration of environmental values than has been achieved using command-and-control regulation.

Whilst not wishing to conflate these two important issues, the second question is probably unanswerable until more case history on NEPI deployment in the member states is available. Equally, the theoretical pros and cons of NEPIs have been meticulously rehearsed in the literature and their restatement here would only serve as a distraction. For a favourable overview of NEPIs, the reader is directed to Pearce and Barbier (2000) and for counter-arguments to Beder (1996). The approach taken in this review is to focus principally on whether NEPIs are causing convergence or divergence of member state environmental policies, pointing out, where appropriate, the environmental benefits and disadvantages of current trends.

6.3.1 *NEPIs: A Source of Convergence or Divergence of Member State Environmental Policies?*

The monitoring of the application of EU environmental policies has proved to be one of the Commission's most daunting tasks. By any interpretation, the criteria laid down for this purpose are extremely malleable and in many cases both plaintiff and defendant are able to cite aspects of EU law to support their case (Heyvaert, 2001). In many respects this situation should not alter significantly with NEPIs, as all directives require national governments to devise implementing mechanisms and there is no logical reason to assume that these have necessarily become more convoluted since the advent of NEPIs. The complexity of implementing command-and-control standards is clearly shown by the Drinking Water Directive (80/778/EC), where member states were required to develop the following myriad of implementing procedures to transpose a single piece of EU legislation:

- Monitoring, sampling, and testing of drinking water
- Identification of supplies that did not comply with mandatory standards
- Planning of measures to remove pollution from water sources, either by treatment or removal of the source from use
- Initiation of capital investment strategies to cope with the expenditure needed to meet the above requirements
- Assessment of problems caused by nitrates and pesticides from agricultural sources as well as lead from lead distribution systems
- Assessment of costs to users
- Development of staff training necessary to meet EU requirements

(Barnes and Barnes, 1999: 108)

It is also perfectly conceivable that as the paradigm of environmental economics becomes more popular and NEPIs gain further credence, there may be a general convergence of implementation methodologies in the member states as governments adopt policies that have proven successful in other countries. One case in point is Germany's Green Dot scheme, which has been implemented in modified form by several member states, including France, Spain, Belgium, Portugal and Luxembourg, as well as by countries outside the EU (DSD, 1998c). Another instance of policy convergence has been the formation of ecological tax reform (ETR) commissions in Belgium, Denmark, Finland, Sweden and Germany (Gee and von Weisäcker, 1994; Gago and Labandeira, 2000).

That said, evidence from the Packaging Waste Directive suggests that NEPIs can also reinforce existing differences in the way member states implement EU environmental policies. Despite commending the general progress made to reach the Directive's targets, the Commission recently expressed concerns that the implementation methods used by the member states were leading to a greater diversity of recovery and recycling standards than was provided for in the Directive's derogations (Cooper, 2000). Whilst the practicalities of designing effective packaging waste management systems will obviously vary between countries, the economic instruments and negotiated agreements deployed in the UK and Germany were determined as much by national and political preferences as by any 'objective' evaluation of possible solutions to the problem of packaging waste (Bailey, 1999a). In general terms, governments concerned about the economic impacts of environmental policy may tend to introduce a larger number of flexible mechanisms than those with stronger environmental policies. In the case of the Packaging Directive, the UK administration government adopted numerous flexible mechanisms to alleviate the costs of implementing EU recycling targets. Whilst analogous mechanisms have been utilised in Germany – with the notable exception of tradable permits – the UK version of producer responsibility is noticeably more industry-friendly than its Germany counterpart, the DSD. Similarly, PRN costs remain a fraction of Green Dot charges, reflecting the different philosophies underpinning UK and German packaging waste policy.

Another area where similar distinctions can be identified is climate change policy. When the UK government first publicised plans for its Climate Change Levy (CCL), the annual revenue from the tax was projected to be approximately £1.75 billion. However, following intense industry lobbying, the Chancellor announced that the rate of the levy would be lowered to produce an expected income of £1 billion in the first year (DETR, 1999e; HM Treasury, 1999; HM Customs and Excise, 2000). Other secondary mechanisms have also been established to reduce the financial burden of the CCL. The first is the introduction of Climate Change Levy Agreements (CCLAs) for energy-intensive sectors, which allow an 80 per cent reduction in the CCL in exchange for negotiated reductions in energy consumption. The rationale for CCLAs is that energy-intensive sectors would be excessively penalised by the CCL and, therefore, should be granted financial concessions provided they adopt realistic but ambitious energy efficiency programmes. The second flexible instrument introduced in the UK is the carbon-trading scheme, which is again designed to help industries mitigate the cost of improving energy efficiency.

As with the Packaging Waste Directive, Germany has incorporated fewer flexible mechanisms into its climate strategy. In 1999, the government introduced or increased eco-taxes across a range of energy products and envisages further rises and reductions in concessions to stimulate the development of new technologies and more economical energy consumption (*BMU*, 2000; Kohlhaas, 2000). The only other flexible mechanism related to climate change in Germany is a broad agreement between the federal government and 19 industry organisations and trade associations. In the first agreement in 1995 German industry set itself the objective, on a voluntary basis, of reducing carbon dioxide emissions by 20 per cent below 1990 levels by 2005. This was extended in 2000, setting an emissions-reduction target for 2008 of 25 per cent with a view to achieving 35 per cent by 2012 (*Bundesverband der Deutschen Industrie*, 2000). In an interesting contrast with CCLAs, reducing the cost of implementing the Kyoto Protocol does not appear to be a primary objective of the voluntary declaration, as the targets exceed those agreed by Germany under the EU's Burden Sharing Agreement. The agreement also states that the federal government and business will decide jointly on the introduction of any further flexible mechanisms. Whilst these examples do not provide conclusive evidence of a dichotomy between the patterns of NEPI adoption in environmental leader and laggard states, they hint that flexible instruments can reinforce longstanding differences in the way member states prioritise and expedite environmental policies.

6.3.2 Removing Forces for Convergence: the Demise of the Push-Pull Dynamic

Although it is not entirely clear whether NEPIs will make the Commission's monitoring role more irksome than has been the case with command-and-control regulation, one of the casualties of NEPIs may be the 'push-pull' dynamic that has characterised much of EU environmental policy over the past thirty years. To re-cap briefly from Chapter two, his process occurs where environmental leader states introduce new policies which have the potential to impede the free movement of goods between member states (Sbragia, 1996). Examples of such restrictions include the Danish ban on non-renewable drinks containers and the provision in the Packaging Ordinance specifying that reusable beverage containers should command a 72 per cent share of the German drinks market, both of which disproportionately increased costs for non-domestic producers. Faced with the prospect of uneven trade conditions between member states, the Commission has three options. First, it can propose harmonising legislation if it feels the measure is essential and promotes the objectives of the EAPs. Second, it can seek prosecution of the leader state for a breach of Article 28 of the Treaty. Finally, it can rule that the policy is an indispensable and proportional response to a health or environmental problem without proposing EU legislation if the trade impediment is deemed to be fairly minor (Heyvaert, 2001).

Assuming a leader state can defend the environmental rationale of its legislation, the policy may be 'pushed' onto the EU agenda in the form of a proposal for a new directive to eliminate the trade barrier. This is then debated by the Council and European Parliament, where objections may be lodged by more environmentally laggard states on the grounds that the planned legislation is either scientifically unjustified or economically damaging to implement. Notwithstanding the advent of qualified majority voting, there remain strong political pressures to achieve consensus in Council on high profile issues because national ministers do not wish to be seen undermining the broader process of European integration (Collins and Earnshaw, 1993). Assuming agreement can be reached, the directive's provisions are usually weakened in order to secure a political accommodation between leader and laggard states in the Council and between the Council and Parliament. A process of policy normalisation then generally occurs with the transposition of the directive into national law. Whilst the push-pull dynamic tends to operate in a rather piecemeal fashion and has encouraged a bargaining rather than problem-focused mentality in environmental policy, the need to uphold the internal market has contributed significantly towards the raising of environmental standards across the EU (Weale, 1996).

For a variety of reasons, the push-pull dynamic will probably continue to operate in some form despite the current trend towards NEPIs. First, directives remain the preferred method for introducing EU environmental policies because they offer member states considerable autonomy over how standards are to be achieved (Jordan, 1999). The likelihood is that standards-based regulation will remain an essential part of EU environmental policy, maintaining the trade tensions that catalyse the push-pull dynamic. Second, member states' apparent keenness to deploy NEPIs as a complement to, rather than as a replacement for, command-and-control is likely also to perpetuate the push-pull dynamic. Third, although the trade effects of NEPIs may be marginal and difficult to determine, they could still have potential repercussions for EU trade unless price incentives or trading conditions are harmonised across all member states.

The traditional benchmark used by the Commission to assess environmental policies – EU or national – has nonetheless been legal standards. Where these are removed or subordinated by NEPIs, the grounds on which the Commission can dispute national policies may be diminished or at least made less easy to discern. This trend may be further exacerbated by an increase in the use of framework directives in environmental policy. Framework directives were introduced in 1992 to resolve some of the inconsistencies caused by the rather *ad hoc* accretion of environmental legislation since the beginning of the EAPs. They are also seen as a device for coping with the increased diversity of the EU resulting from the accession of new member states. Finally, they are intended to allay fears of over-federalisation by giving national authorities more say over how the detailed objectives of EU environmental policies are defined (Lowe and Ward, 1998b). Framework directives achieve these aims by setting out general principles, procedures, and requirements rather than detailed standards. However, more detailed 'daughter' directives may be introduced within the scope of a framework directive where they are considered necessary to protect a particular aspect of environmental quality (CEC, 2000d). Whilst framework directives had only been adopted for the air, waste, and water sectors by the end of the 1990s, Barnes and Barnes (1999) suggest they will continue to gain in prominence as the scope and influence of EU environmental policy continues to broaden. Erosion of the push-pull dynamic would therefore be most pronounced for environmental policies enacted through framework directives and/or NEPIs, particularly purer forms of market-based mechanism.

One example of where differentiated national NEPIs has not triggered the push-pull dynamic is again the case of energy taxation (Ekins and Speck, 2000). The Commission tabled numerous proposals during the late 1980s and 1990s for a common EU carbon/energy tax to reduce the emission of greenhouse gases implicated in global climate change. At the same time, the introduction of an EU carbon/energy tax was seen as preferential to national taxes because it reduced the potential for trade distortions caused by uneven fuel duties. The proposal therefore exhibited many of the prerequisites of the push-pull dynamic, not least the necessity for trade harmonisation and a keenness amongst some member states – notably the Netherlands, Denmark and Germany – to introduce national energy taxes (Zito, 2000). However, despite repeatedly being placed on the Council agenda, a common position on the EU carbon/energy tax proposal has not materialised. This impasse was caused principally by the requirement that all measures of a fiscal nature must receive unanimous support in Council[9]. This proved difficult to achieve as ministers from less affluent member states – especially Greece, Ireland, Portugal and Spain – argued that common carbon taxes would impose a unmanageable burden on their economies, whilst Britain objected in principle to the idea of EU intervention in national fiscal policies (Zito, 2000).

In an effort to revive negotiations, in 1997 the Commission tabled a more modest proposal to establish a framework for the taxation of energy products (CEC, 1997a). The objectives of this draft directive include the reduction of disparities between national energy taxes, the restructuring of national energy-tax systems, and the extension of minimum excise duties on mineral oils through three increases over a five-year period (Ekins and Speck, 2000). Despite the greater flexibility of these suggestions, the European Environment Agency (EEA) reported in 2000 that little progress had still been made in agreeing common energy tax rules, again because of a lack of Council support (EEA, 2000). Member states have instead opted to address the climate change issue by introducing or extending national energy-tax schemes. This has led to an uneven picture of energy taxes across the EU member states (Table 6.1).

In terms of tradable permits, the Commission would again struggle to use the UK carbon trading system as legitimation for a European scheme except by publicising its merits, since the British scheme does not stipulate environmental standards that might restrict access to the UK market. Instead the Commission has been forced to rely on the environmental and economic logic of supranational carbon trading to defend its proposals rather than being able to invoke the tried-and-tested trade justification used to promote the Packaging Waste Directive. This is not to suggest that the Commission's arguments for EU carbon trading are insubstantial but merely that the trade card has proven a convenient and effective device for championing national measures it wishes to convert into EU legislation. Finally, voluntary agreements could further erode the push-pull dynamic where national regulators and industry negotiate non-binding environmental targets, as the Commission would be unable to prove a trade case for policy harmonisation (Nunan, 1999; Börkey and Lévêque, 2000).

There are sound reasons to conclude, therefore, that NEPIs provide leader and laggard governments with greater scope to pursue their preferred environmental strategies without necessarily creating the trade tensions required for the operation of the push-pull dynamic. The corollary of this argument is that further divergence between the environmental outcomes achieved in the member states will result from further increases in the uptake of NEPIs. Against this, legislative harmonisation in the form of a continuing stream of environmental directives and regulations is likely to maintain

Table 6.1   Energy taxes in selected EU states, June 2002

	Electricity kWh^{-1} unless stated			Coal $tonne^{-1}$		Heavy fuel oil $tonne^{-1}$	Unleaded petrol $tonne^{-1}$	Diesel $tonne^{-1}$
	General	Domestic	Industrial	Coal	Lignite			
Austria	0.015						0.407	0.282
Belgium	0.001					6.2-18.6	0.494	0.290
Denmark	0.013	0.067	0.076	32.49	23.89	37.6	0.542	0.369
Finland		0.007	0.004	41.40	14.40	54.0	0.561	0.300
France	0.013					16.8-23.2	0.587	0.389
Germany		0.002	0.003			57.8	0.563	0.379
Italy						60-130	0.540	n/a
Netherlands	0.014-0.058[a]			11.22		155.1	0.592	0.326
Spain	4.864%		0.015[b]			13.4	0.372	0.270
Sweden		0.002					0.499	0.337
UK	0.069			18.80	18.80		0.786	0.738

a: depending upon electricity usage
b: excludes manufacturing industries

The figures represent only the nominal tax rates for chosen tax bases. In many cases there are exemptions and refund mechanisms related to the tax that can have an impact on effective tax rates. Figures for natural gas are not presented because of variations in units of measurement.

Source: OECD (2002)

constraints on how far diversity pervades the implementation of EU environmental policy.

Equally, a reduction in the push-pull dynamic should not automatically be seen as damaging to the broader ambition to promote sustainable development, if it enables leader states to pursue ambitious agendas whilst allowing others to develop policies at a more measured pace consistent with their political and economic circumstances. As Lowe and Ward (1998b) note, the concept of a more flexible EU environmental policy has been stimulated by a recognition of the need to avoid stagnation in the face of such challenges as enlargement and unease amongst some member states over further increases in the pooling of sovereignty. Moreover, if more flexible environmental policy proves to be compatible with the demands of the internal market and NEPIs help to fulfil this ambition, then further experimentation with new policy instruments by leader states may lead to notable successes, which more conservative states may then be eager to emulate.

6.4   Conclusion

Finding a term that adequately captures the character of European integration has proven a challenging task for politicians and analysts alike. Indeed, de Tocqueville's contention (in Höreth, 1999: 249) that the EU is a new form of governance, the word to describe which does not yet exist, may well be the most apt conclusion. As with many aspects of the European project, EU environmental policy remains lodged somewhere between federalism and intergovernmentalism and no single mode of governance has consistently dominated all others. Nor would a wholly federal outcome necessarily be desirable considering the diverse challenges faced by the member states in respect of environmental policy (Jordan, 1999).

Whether NEPIs will fundamentally alter the current balance between the confederal and intergovernmental personas of EU environmental policy remains to be seen. However, it seems unlikely that increased federalisation of policy implementation will become a major component of the EU's environmental strategy, at least in the short to medium term. The practical and constitutional difficulties of developing NEPIs at a supranational level were clearly exposed by the EU carbon/energy tax proposal. Experimentation with NEPIs is therefore likely to remain focused predominantly at a national level though, as the ACEA-JAMA-KAMA agreements demonstrate, the development of further EU NEPIs cannot be entirely ruled out where there are clear justifications for supranational policies.

In evaluating the degree to which NEPIs are liable to promote policy convergence or divergence, it is important to remember that the priorities set out by the Treaty and EAPs will continue to stress the importance of collective action to protect the environment. Equally, the apparent universal preference for command-and-control directives and regulations as the principal instruments of European environmental policy will be critical in maintaining cohesion. Finally, the extent of future diversity also depends on member states' appetite to engage with active policy learning and transfer, as new variations and applications of NEPIs are found. There is nevertheless a distinct possibility that NEPIs will prompt greater diversity in the implementation of EU environmental policy. Leader governments are likely to remain under pressure from their electorates and NGOs to find innovative ways of integrating environmental concerns into economic and social planning – though this enthusiasm may be tempered where this entails significant new taxation – whilst laggard states will probably continue to implement EU law in a less ambitious manner. In both cases, NEPIs provide immense scope for the modification of EU environmental policies during practical implementation. The outcome of this increased flexibility is likely to be continued diversity between the effective environmental standards achieved in leader and laggard states despite the ongoing process of legislative harmonisation.

However, this divergence in no way signifies a return to more intergovernmental modes of decision-making, since the EU's right to define the agenda of environmental policy in the member states has in no way been diminished by the advent of NEPIs. More likely, similar tensions to those currently exhibited with command-and-control policy implementation will persist but be magnified as the range of policies used to protect the environment expands. NEPIs also do nothing to reduce other causes of the implementation deficit in the EU, such as the difficulties of reconciling EU and national law, monitoring practical implementation and the cumbersome nature of infringement proceedings.

However, if NEPIs are to contribute significantly to the effective implementation of EU environmental policies in the member states, clearer guidelines may be needed to ensure that conflicts between nationally applied NEPIs and the internal market do not become more widespread and inhibit innovation in environmental policy. The general principles currently used to adjudicate on trade disputes have proven extremely malleable and led to an accretion of rulings that leaves many uncertainties as to what member states may or may not do in relation to the design and implementation of environmental policies. By developing more detailed procedures for nationally implemented NEPIs, for example, under the auspices of framework directives, the EU can assist in promoting environmental policies that are sensitive to national contexts and political priorities whilst continuing the process of reforming the command-and-control approach towards environmental policy. This is no straightforward task, however. The devil is always in the detail and managing the transition to sustainable development is likely to be a major test for the overseers of European integration. To conclude this review, the final chapter assesses future challenges for the deployment of NEPIs in the EU.

Notes

[1] Though there is invariably some degree of leakage between nominal environmental standards – those set by legislation – and the effective standards achieved in practice (see Chapter two).

[2] However, Ekins and Speck (2000) argue that there is little reason to expect and no evidence to support the idea that countries which adopt well-formulated MBMs will experience a negative effect on overall long-term economic performance. They suggest that environmental taxes are among the most efficient ways of pursuing many environmental policies, so countries which deploy them judiciously will perform better economically than those that rely on command-and-control.

[3] Community guidelines on state aid for environmental protection (94/C 72/03) were developed in 1993 to clarify where, and to what extent, state aid was permissible (OJEC, 1994b). Aid to support environmental investments is allowed, between 30 and 40 per cent, only if the investments aim to achieve levels of environmental protection significantly higher than those required by mandatory standards. The guidelines also state that exemption from environmental taxes, which is regarded as state aid, is only permitted if this is necessary to prevent domestic companies being disadvantaged compared with competitors in member states where no equivalent measures exist (Debelke and Bergman, 1998).

[4] This occurred with the implementation of the Packaging Waste Directive, where British and German legislation discounted exported packaging from recycling targets but included imports in order to ensure their measures were non-discriminatory.

[5] Other factors causing firms to invest in abatement include pressure from consumers, NGOs or green investors, and the company's general commitment to environmental protection. The first two are again principally financial motivations, whilst the latter reflects concerns not necessarily linked to the cost minimisation or profit maximisation motive of the firm (Pearce and Barbier, 2000). One inducement for corporate environmental commitment has been accreditation to environmental management standards such as ISO14001. This is often classified as a form of voluntary agreement and, again, may produce financial benefits for accredited businesses.

[6] The Landfill Directive requires that the amount of biodegradable municipal waste going to landfills be reduced to 75 per cent of the total amount (by weight) produced in 1995 within five years of the Directive entering into force, to 50 per cent within eight years, and to 35 per cent within fifteen years (DETR, 2000a). There are already numerous examples where waste in the UK is transported between local authority areas, and sometimes considerable distances, because of shortages in landfill space in the area where the waste is generated (personal interviews with waste management companies).

[7] A contrast can be drawn here with the so-called 'California effect' in the USA, where America's most economically powerful state has sought to make the adoption of its high environmental standards a pre-condition of trade with states that have less stringent environmental laws (Vogel, 1997).

[8] Some prominent NEPIs have been developed at a European level, however. Two of the most longstanding are the Eco-Management and Audit Scheme and the Eco-Label Scheme (Barnes and Barnes, 1999). Another is the agreements between the Commission and European, Japanese and Korean vehicle manufacturers to reduce carbon emissions from new private cars, the so-called ACEA-JAMA-KAMA agreements (Keay-Bright, 2000). The justifications for an EU agreement in this area are twofold. First, the emission of greenhouse gases from the transport sector is a trans-national problem and would appear to warrant a supranational response. Second, the automotive industry consists primarily of trans-national companies. Consequently, an EU agreement encourages motor manufacturers to reduce vehicle emissions rather than relocating to member states with more lax emissions policies. In order to meet the subsidiarity tests set out in the 1992 Edinburgh Council, the Commission is nevertheless required to justify any new policies in terms of their trans-national effects (spillover), free trade implications, or the value added by an EU initiative (Jordan, 2000). Whilst there was a general consensus in favour of the ACEA-JAMA-KAMA agreements, concurrence for more extensive EU involvement in the design and deployment of NEPIs would be difficult to obtain.

[9] As noted earlier, under Article 93 unanimous voting is also applied for measures affecting town and country planning, quantitative management of water resources (including policies impacting on the availability of those resources), land use (with the exception of waste management), and member states' choice between different energy resources and the general structure of their energy supply (OJEC, 2001: 19).

Chapter 7

Conclusions and Prospects for the Future

7.1 Introduction

The central aim of this book has been to examine the extent to which NEPIs have contributed towards the effective implementation of environmental policies in the European Union. In pursuit of this aim, the Packaging Waste Directive has been used to highlight broader issues concerning the translation of economic theory on environmental policies into practical action in the EU member states. Three key objectives were set. The first was to establish the manner in which NEPIs are being used by the member states to implement EU environmental policy. The second was to investigate the extent to which NEPIs have improved the implementation of EU environmental policy, with particular reference to their impact on industry. The final objective was to assess the effects of EU decision-making and implementation procedures on the performance of NEPIs and, correspondingly, the influence of NEPIs on the harmonisation of national environmental policies. To conclude this discussion, the final chapter reviews the main conclusions of the study and considers the future prospects for NEPIs in the EU.

From the evidence of the preceding chapters, it is apparent that NEPIs have led to major innovations in the way environmental policies are implemented by EU member states. Negotiated agreements have been used in the UK and Germany to establish producer responsibilities and inject greater environmental stewardship into the management of waste. They have also contributed towards a more co-operative relationship between regulators and regulated, giving industries greater autonomy over the definition of environmental targets and how they are to be achieved. In the context of packaging-waste policy, environmental charges have been employed principally as a means of creating flows of funding from waste producers to parties involved in the collection, reprocessing and development of end markets for packaging waste. Finally, the UK government deployed a tradable recycling certificate, the PRN, to facilitate this re-allocation of resources, using competitive trading to mitigate the costs of meeting EU recovery and recycling targets. This has resulted in a steady increase in recycling at lower absolute and relative cost than the more regimented and centrally controlled Dual System.

In some areas, however, it is clear that NEPIs have under-performed against their theoretical potential. Although it is customary for negotiated agreements to contain an element of coercion, the way in which producer responsibilities were constructed in the UK and Germany has led to much discord amongst participating industries. In Britain, there has been protracted wrangling over the apportionment of sector responsibilities whilst in Germany high recycling targets forced the DSD to accept almost any recycling contract it was offered regardless of the financial implications. Both agreements have also witnessed some alarming regulatory lapses. The early years of the DSD were beset by problems caused by the fact that less than adequate provisions were made to fund expansion in reprocessing capacity, whilst difficulties emerged in the UK because reprocessors enjoyed the benefits of receiving PRN income but were not legally obliged to invest in new recycling capacity. Many reprocessors consequently chose to see PRNs simply as a source of windfall profit, leading to a dearth of investment in reprocessing facilities and, more latterly, waste collection. Finally, many companies have chosen to free ride the agreements, taking advantage of lax monitoring and legislative loopholes. The UK and German governments have sought to correct these deficiencies using retrospective command-and-control regulation, including new prohibitions, additional reporting requirements and instructions for enforcement agencies to take a tougher line with free riders, to ensure that the incentives of the market do not undermine those of environmental policy.

Environmental charges in the UK and Germany have proven generally successful in raising funds for recycling infrastructure – though less so in persuading the UK reprocessing sector actually to invest in new facilities – but have not been a major factor behind corporate decisions to reduce, re-use and recycle their packaging waste. Such changes in behaviour have instead been brought about primarily through command-and-control measures. Whilst the Green Dot and PRNs were both designed as cost-covering charges rather than incentive taxes, the fact that significantly higher charges in Germany have not produced a discernibly greater incentive for waste reduction may have more general implications for the use of economic instruments to regulate some environmental problems. In this instance, the lack of abatement incentives emerging from the charges is explained principally by the high demand price-inelasticity of packaging and the fact that both instruments reflect the investment needs of national recycling industries rather than the marginal environmental cost of packaging production and waste management. As similar factors affect many pollution markets, there is little basis for policy-makers to assume, *a priori*, that cost-covering charges will produce an incidental abatement incentive except by pure coincidence. Equally, if taxes or charges on price-inelastic commodities are set at punitive rates in order to induce an incentive effect, this may undermine the economic efficiency of the policy and, hence, a principal justification for the introduction of an economic instrument. Decision-makers must therefore clearly identify the characteristics of each pollution market and the main objectives of economic instrument in advance of implementation in order to select the most appropriate combination of policy instruments.

The UK government's scheme of tradable recycling certificates was the most experimental of the NEPIs used to implement the Packaging Waste Directive; as such, it is not surprising that the PRN system has experienced teething troubles. The scheme's main innovation was to introduce trading between producers of packaging waste and providers of abatement services. However, a general level of mistrust has developed between the two sectors, with producers accusing reprocessors of manipulating PRN prices and reprocessors claiming that producers and compliance schemes were not providing sufficient guarantees to justify investments in new capacity. These problems generally stem from the inadequacy of the rules governing the use of PRN revenue. In short, the UK government placed excessive in the idea that profit-led trading would produce natural synergies with the objective of improved environmental protection. Once additional rules were introduced to stipulate guidelines on the utilisation of PRN proceeds, the system has by and large attained greater stability. This experience again highlights the general lesson that although NEPIs have the potential to produce cost-effective improvements in environmental protection, sight should not be lost of the fact that the natural tendency of free markets is to operate for private benefit rather than wider social and environmental objectives. It is therefore essential that potential conflicts between these prerogatives are identified and removed during policy design. In many cases, this means the deployment of command-and-control measures alongside NEPIs to ensure a continuity of objectives.

The study has also revealed some intriguing issues concerning the role of NEPIs within the EU. First, although tensions undoubtedly exist between the desire to maintain flexibility in the implementation of EU environmental policies and the need for harmonised rules within the Single Market, initial indications are that the level of conflict between NEPIs and the internal market is not especially great. Although disparities in national environmental taxes have emerged, these form only a minor component of the factors affecting the competitiveness of firms within the European economy. Moreover, the EU has moved progressively away from automatically prohibiting national environmental policies that impede free trade towards assessing restrictions in terms of their indispensability, proportionality and, more recently, the extent to which they promote a systematic substitution of environmentally damaging products and processes. Nonetheless, for some NEPIs, more formal means of preventing trade restrictions, such as border tax adjustments, may be required to protect competition in the internal market.

Finally, as has generally been the case with the implementation of EU environmental policies, the design and deployment of NEPIs continues to be dominated by national governments, with only faltering progress being made towards the development of EU-level NEPIs. EU intervention in the implementation of environmental policy remains an extremely contentious issue and most member states have jealously guarded their right to determine the way in which EU standards are applied. This approach has many benefits, not least allowing the design of NEPIs to be overseen by authorities best placed to judge local contexts and circumstances. This may prove to be a major factor in reducing the implementation deficit in EU environmental policy. At the same time, the flexible nature of NEPIs increases the likelihood of further divergence in *de facto* environmental policies in the member states, which may in turn prompt more serious conflicts between the environmental programme and the internal market.

To conclude the discussion the remainder of this chapter examines the future prospects for NEPIs in the EU. Specifically, it focuses on the challenges likely to emerge from the introduction of the European Single Currency, the Euro, and the imminent accession of new member states to the EU (Blacksell, 1998; Barnes and Barnes, 1999). The next two sections review the implications of these events for the EU environmental programme and, in particular, the deployment of NEPIs.

7.2 NEPIs and EU Monetary Union

With the adoption of the Single Market Programme in 1985, the Commission increasingly took the view that the full potential of the internal market could not be realised as long as currency conversions and exchange rate fluctuations continued to impose uncertainties and transaction costs on importers and exporters. Many economists also argued that the free movement of capital, exchange-rate stability and independent monetary policies constituted an 'impossible triangle' that was fundamentally irreconcilable in the long term. In June 1988 the Hanover European Council set up a committee to study economic and monetary union, chaired by Jacques Delors, then President of the European Commission. Its report, submitted in April 1989, proposed the introduction of economic and monetary union (EMU) in three stages; the full liberalisation of capital movements, the development of binding rules on the size and financing of national budget deficits and the establishment of a new, independent institution responsible for the Union's monetary policy (Pinder, 1996). Following extensive preparations and a formal commitment to EMU in the Maastricht Treaty, the Euro replaced national currencies in twelve of the member states from 1st January 2002. However, three member states, the UK, Denmark and Sweden, opted not to join the Euro-zone immediately because of concerns over sovereignty and the economic wisdom of integrating diverse European economies in such an irrevocable manner.

Simplifying slightly, the purpose of the Euro is to promote EMU through the co-ordination of monetary policies and the expansion of intra-Community trade. It is also suggested that the Euro might ultimately lead to further reforms that result in the EU becoming a more wholly federal system of governance (Pinder, 1996). In terms of the environmental impacts of the Euro, both positive and negative consequences can be identified. For example, the single currency has the potential to increase the specialisation of production in the member states which, if based on the principle of comparative advantage, could encourage greater efficiency of resource use (Bhagwati, 1993). Conversely, trade specialisation could escalate transport emissions, reduce local employment choices, and cause a dangerous spatial dislocation between centres of consumption in the wealthier Northern states and the environmental impact of that consumption, as production and pollution shifts towards regions with lower employment and land costs (Daly, 1993).

One of the more immediate effects of the single currency on NEPIs will undoubtedly be a greater transparency in the design of economic instruments, as governments, businesses and citizens can more readily compare national variations in environmental taxes. If major discrepancies in fuel duties, for example, can be identified, this might increase the pressure for harmonisation, especially if consumers in countries with higher excise taxes see that they are being penalised more heavily than other areas of the Single Market. The Commission and national governments should also be able to identify at an earlier stage when a member state is creating pollution hot spots by maintaining economically favourable rates of environmental tax. However, such pressure for convergence would not automatically lead to an upward shift in environmental taxes and charges, as the prime environmental leader states are not necessarily those with the highest rates of eco-taxes. The EEA (2000) notes, for instance, that Germany derived a lower proportion of total revenues from environmental taxes in 1997 than the EU average (approximately 5.25 per cent). Although this situation has changed somewhat with the formation of the German Ecological Tax Reform Commission in 1999, there is little indication of the 'traditional' divide between environmental leader and laggard states in respect of environmental taxes. In 1997 environmental taxes exceeded eight per cent of total tax revenue in seven member states, including two 'leaders' (Denmark and the Netherlands), one 'pragmatist' (the UK), and four countries with a reputation for having relatively lax environmental policies (Portugal, Greece, Ireland, and Italy) (EEA, 2000: 24).

The political and practical obstacles to the harmonisation of environmental taxes and charges nonetheless remain formidable, not least because this would require further extensions to the competencies of the EU – a move which would be anathema to some member states. The alternative might be to introduce formal controls on economic instruments to stave off serious conflicts between environmental and trade policy. One such device, border tax adjustments (BTAs), was reviewed in Chapter six. A variation on this to prevent harmonised taxes becoming overly regressive would be to approximate environmental taxes on the basis of purchasing power parity between richer and poorer member states. However, political objections to EU interference in national fiscal policy would still be difficult to surmount.

The principal effect of EMU on tradable permit schemes is its potential to facilitate international trading by reducing currency conversions and other transaction costs. The Commission has already tabled proposals for a scheme for greenhouse gases and similar economies of scale could be gained from permit trading between member states for other pollutants (Capros and Mantos, 2000). Whilst this could help to reduce the cost of environmental policy, unregulated trading across the EU may exacerbate local environmental impacts caused by regional concentrations of emissions. As with all permit schemes, therefore, rules are required to ensure that equitable environmental standards are maintained (Atkinson and Tietenberg, 1991). One such mechanism, suggested in Chapter six, is the use of framework directives to set out broad pollution standards, which member states could then administrate using permit trading. EU-wide schemes may also produce secondary economic and environmental benefits by helping to transfer funds and technologies to less affluent member states. In the case of packaging waste, UK membership of the Euro might make it economic for compliance schemes to obtain more PERNs (PRNs produced by non-domestic reprocessors) from states with lower recycling targets, such as Ireland, Portugal and Greece, whilst simultaneously providing financial support for their recycling industries. Overall, therefore, the Euro may help to overcome some of the difficulties that have militated against the widespread adoption of this policy instrument.

EMU is unlikely to bring significant changes in the design and implementation of other NEPIs, such as voluntary environmental agreements and informational devices, as neither involves the creation of a defined fiscal instrument. However, one final issue in relation to the impact of the Euro on NEPIs is noteworthy. Although the Euro-zone consists of twelve member states, the refusal of UK, Denmark and Sweden to join in the first wave may heighten political tensions concerning the negotiation and implementation of EU environmental policies. Having taken the step of agreeing to a single currency, it is by no means inconceivable that those within the Euro-zone might see the adoption of common NEPIs, albeit with the customary derogations and opt-ups, as a logical extension to this stage of European integration. This may further isolate the UK in environmental decision-making, especially if it refuses to countenance EU involvement in fiscal policies and exercises its veto under Article 93 to frustrate the ambitions of more integration-minded member states. Equally, if Britain and the other Euro abstainees become marginalised from international permit trading schemes, this may have a detrimental economic impact, both nationally and for the EU scheme[1].

It should be noted that any predictions about the effects of the Euro on environmental policy or the design and operation of NEPIs are highly speculative until such time as the broader economic ramifications of the single currency become more apparent. Nonetheless, this brief review has sought to make initial predictions based on the known attributes of NEPIs and political issues relevant to EU environmental policy and it is hoped that this initial exploration will encourage further debate on this issue. Overall, the Euro appears to offer some benefits for environmental protection in the EU by facilitating the co-ordination of national policy instruments but, by the same token, integration of this nature would require further reforms of the protocols used to negotiate EU environmental policies, particularly the current unanimity rules for fiscal measures, if progress is not to be inhibited by concerns over creeping federalisation. This might involve provisions allowing member states to harmonise economic instruments on a voluntary basis so as to allow those that wish to proceed on this basis the right to do so. As this could aggravate tensions between trade and environmental policy, accompanying measures, such as BTAs, might be required to guarantee that trade blocs did not evolve within the internal market. Safeguards would also be necessary to ensure that member states participating in such schemes do not lose their essential and democratic right to tailor EU environmental policies to fit national requirements and exigencies.

7.3   NEPIs and EU Enlargement

There is little doubt that future enlargements to the EU will bring important changes to the political landscape of environmental policy, especially as ten of the thirteen states currently applying for EU membership are from Central and Eastern Europe (CEE) (CEC, 1997b)[2]. With the collapse of state socialism in the late 1980s and early 1990s, the CEE states have embarked upon a laborious transition to Western political and economic philosophies. As Klarer and Francis (1997: 1) point out:

> Entire economic, political and social systems are being re-built ... New democracies are being built through 'top-down' legislation and policy and the 'bottom-up' demands and participation of individual citizens and independent organizations ... Economic recovery and growth based on free market economies have become the pre-eminent priorities for most CEE countries.

As part of this transition, the majority of CEE governments have also expressed a keen desire to become re-integrated into the mainstream of European affairs and many have made membership of the EU a key priority (Blacksell, 1998; Saiko, 1998). Responding to these overtures, the Commission set out three criteria which the CEE states must fulfil before joining the EU at the European Council in Copenhagen in 1993 (CEC, 1993):

*   The stability of institutions guaranteeing democracy, the rule of law, human rights and respect for and protection of minorities;
*   The existence of a functioning market economy as well as the capacity to cope with competitive pressure and market forces within the Union;
*   The ability to take on the obligations of membership, including adherence to the aims of political, economic and monetary union.

These criteria present the CEE states with profound challenges across the entire policy spectrum as, for several decades, many CEE societies lacked even basic familiarity with the principles of democracy and market economies (Klarer and Francis, 1997). However, alignment with EU environmental policy would appear to present particularly major problems for the CEE states. Almost without exception, these countries suffered severe environmental mismanagement during the socialist era, caused by myopic central planning, chronic under-investment in new technologies and abatement facilities, and a culture of administrative secrecy, which kept many environmental problems hidden from the public view (Manser, 1993; Saiko, 1998; 2000). In order to develop a structured approach towards the alignment of environmental policies, the Commission has identified the following as the main tasks facing the CEE countries (CEC, 1997b):

- The need to transpose the EU environment *acquis* into national legislation, based on a strategic approach towards the protection of the environment and human health;
- The creation or strengthening of administrative structures to oversee the implementation, monitoring and enforcement of environmental standards;
- The availability of financial resources and expertise to support the remedial works and infrastructure developments required for an effective system of environmental protection.

The Commission's report to the European Council also identified sector-specific challenges that required most urgent attention, including air quality, waste management, water quality, industrial pollution control and risk management, and nuclear safety and radiation protection (CEC, 1997b). A full review of progress towards these aims is not possible here but three issues concerning the CEE accessions are of particular relevance to the main themes of this book. The first concerns preparations for the harmonisation of CEE environmental policies with those of the EU, the second, the effects of enlargement on EU environmental decision-making procedures, and the third, the role of NEPIs in the environmental rehabilitation of the CEE states. These are now examined in turn.

7.3.1 Adopting the Environmental Acquis in the CEE Countries

The first stage in the preparations for enlargement is for CEE states to align their laws with those of the EU. The initial view of the European Commission, expressed at the Environment Council in 1996, was that this should include all aspects of the environmental *acquis* before accession takes place (Mayhew, 1998). Progress against towards this target has proved slow, however, and it is estimated that full harmonisation could take in excess of ten years in some cases because of the sheer scale of the work and the costs involved (Turner, 1997, cited in Barnes and Barnes, 1999; Saiko, 1998). According to the Commission, even Slovenia faces immense practical and institutional challenges in conforming to EU environmental standards despite enjoying one of the most prosperous economies in Eastern Europe (CEC, 1999b). At the Cardiff European Council Summit in 1998, it was recognised that it would not be possible for the candidate states to introduce all EU environmental legislation before accession and that a more pragmatic approach was needed to avoid serious delays. The Commission consequently suggested that derogations should be negotiated on a country-by-country basis, contingent on each candidate country making demonstrable and consistent progress towards alignment across the full range of the environmental *acquis* (Barnes and Barnes, 1999).

The Commission publishes annual strategy documents and more detailed regular reports for each country in order to review progress towards accession (CEC, 2001c). In its 2001 strategy document, the Commission commented on the positive efforts and substantial progress being made by all CEE states in harmonising legal standards but also noted that implementation and the cost of alignment were proving major challenges in some cases. Levels of monitoring and enforcement were also considered to be problematic, as was the availability of finance and expertise. At the time of writing, the Commission concluded that none of the candidate countries had fulfilled all the environmental criteria required for accession.

The outline strategy presented at the Cardiff Summit also proposed that the majority of the funding for environmental approximation should come from the applicant states themselves with the EU making only a partial contribution through such financial instruments as PHARE (Barnes and Barnes, 1999)[3]. From 2000, PHARE will provide Euro 1.56 billion annually for economic and social development and industrial reconstruction, including municipal infrastructure, industry-related investments in clean up, and reductions in hazardous waste and emissions. The Commission has also proposed a package of new instruments to assist with the costs of accession. Among these the ISPA (*Instrument structuralle pour le pre d'adhesion* (Pre-accession structural assistance)) is the most significant, providing Euro 7 billion over seven years from 2000, half of which will be devoted to environmental infrastructure investment. EU funds nonetheless only cover approximately four per cent of the investment needs of the CEE states, though a number of other bodies have established funds to aid environmental and economic transition in the CEE nations, including the European Investment Bank and the Global Environmental Facility (Klarer and Francis, 1997). Finally, although the cost of meeting the *acquis* in the candidate countries is estimated to be Euro 120 billion, the Commission calculates that the benefits of harmonisation will amount to Euro 140 billion over ten years and therefore suggests that short-term investments will yield long-term financial as well as environmental dividends (CEC, 1997b).

7.3.2 *Environmental Decision-making in an Enlarged EU*

Aside from the practical challenges posed by enlargement, each wave of accessions would logically appear to have the potential to change the political dynamics of EU environmental decision-making (Barnes and Barnes, 1999). If true, the accessions of the 1980s, which brought Greece, Spain and Portugal into the Community, should have shifted the political pendulum towards more cautious environmental policy, whilst the 1995 enlargement (incorporating Austria, Finland and Sweden) ought to have increased the influence of the 'green' lobby in key institutions. However, there is little evidence that EU environmental policy faltered in the 1980s; indeed, this period saw unprecedented legislative output, the introduction of key integrative policies such as the Environmental Impact Assessment Directive (85/337/EC) and the formal incorporation of environmental policy into the SEA. Equally, Liefferink and Andersen (1998) claim that a green revolution did not occur following the 1995 accessions, as consistent alliances between environmental leader states failed to materialise because of differences in strategic outlook towards environmental policy. Both facts support Weale's (1996) thesis that decision-making power is sufficiently well distributed across the EU institutions so as to prevent a dominant coalition of (either leader or laggard) countries consistently imposing their national styles of regulation onto other member states. Correspondingly, successive enlargements and the formalisation of environmental policy have given rise to a political pre-occupation with subsidiarity and questioning of the EU's authority to regulate in areas of policy previously controlled by national authorities (Jordan, 2000). The consequence of this debate, arguably, has been increased pressure for the Commission to justify new regulations and directives[4].

The fact remains, however, that enlargement increases the number of state actors whose agreement is required for new environmental initiatives to become EU law and here analogies can be drawn with Pressman and Wildavsky's (1984) examination of the implementation difficulties arising from policy-making as a joint decision process. If securing concurrent majorities amongst the current fifteen member states has proven an intricate task, obtaining agreements in an EU comprising twenty or thirty members will be even more difficult (Lowe and Ward, 1998b). Although it should not automatically be assumed that the CEE governments will seek to slow the tempo of EU environmental policy, their potential influence on Council and European Parliament decisions is evident from Table 7.1 (CEC, 2000a), and their key priority has to be the protection of their nascent market economies from excessively rigid environmental standards. Overall, therefore, the CEE

Table 7.1   **Provisional apportionment of voting under an expanded EU**

State	Population (millions)	Population (%)	European Parliament seats	Council votes	Commission
Member States					
Belgium	10.21	1.87	25	5	1
Denmark	5.31	0.97	16	3	1
Germany	82.04	15.04	99	10	2
Greece	10.53	1.93	25	5	1
Spain	39.39	7.22	64	8	2
France	58.97	10.81	87	10	2
Ireland	3.74	0.69	15	3	1
Italy	57.61	10.56	87	10	2
Luxembourg	0.43	0.08	6	2	1
Netherlands	15.76	2.89	31	5	1
Austria	8.08	1.48	21	4	1
Portugal	9.98	1.83	25	5	1
Finland	5.16	0.95	16	3	1
Sweden	8.85	1.62	22	4	1
UK	59.25	10.86	87	10	2
TOTAL	375.36		626	87	20
Candidate countries					
Bulgaria	8.23	1.51	21	4	1
Cyprus	0.75	0.14	6	2	1
Estonia	1.45	0.27	7	3	1
Hungary	10.09	1.85	25	5	1
Latvia	2.44	0.45	10	3	1
Lithuania	3.70	0.68	15	3	1
Malta	0.38	0.07	6	2	1
Poland	38.67	7.09	64	8	2
Czech Republic	10.29	1.89	25	5	1
Romania	22.49	4.12	44	6	1
Slovakia	5.39	0.99	16	3	1
Slovenia	1.98	0.36	9	3	1
Turkey	64.39	11.8	89	10	2
TOTAL	545.56	100	963	144	35

Source: CEC (2000a: 63)

governments are likely to take a largely cautious view towards environmental policy. However, it would be misleading to imply that the CEE accessions will lead to the kind of hiatus experienced in other areas of EU policy between the mid-1970s and the SEA as a result of the OPEC oil shocks and the Luxembourg compromise (Hildebrand, 1993)[5]. It is more probable that unless more flexible decision-making protocols are adopted, enlargement could lead to the re-emergence of lowest-common-denominator-bargaining as a prominent feature of EU environmental policy-making. In the early years of the environmental programme, lowest-common-denominator standards tended to prevail because of the need to obtain unanimous approval in the Council of Ministers for all new environmental policies. On the other hand, the lack of high-level political attention devoted to environmental policy enabled the Environment Directorate to pursue a relatively ambitious legislative programme despite its lack of political influence (Weale, 1999).

Zito (2000) argues that the ethic of lowest-common-denominator bargaining became more contested in the 1980s as a series of environmental entrepreneurs emerged in key positions in the EU institutions to inject greater innovation into environmental policy and move decision-making beyond a focus on short-term and lowest-common-denominator preferences. One such entrepreneur was Carlo Ripa di Meana, whose tenure as Environment Commissioner between 1989 and 1992 was characterised by strong advocacy of more extensive and strategic EU environmental legislation (Lowe and Ward, 1998a). Equally, the European Parliament identified environmental policy as an area where it could extend its influence beyond its traditional place at the periphery of European politics (Weale, 1999). The cause of environmental entrepreneurship was further strengthened by the increasing intensity of many environmental problems, the inclusion of environmental objectives into the SEA, and the introduction of qualified majority voting for many areas of environmental policy (Barnes and Barnes, 1999). Finally, policy entrepreneurship is an integral part of the push-pull dynamic, as actors from environmental leader states become policy advocates in their efforts to persuade other Council representatives to adopt their national regulatory styles.

Zito (2000) nonetheless concedes that environmental entrepreneur-ship and lowest-common-denominator bargaining continue to co-exist in the EU policy process for the simple reason that member states have diverse interests, which still require mediation through the EU process.

> Entrepreneurial coalitions must manipulate the constellation of ideas and interests and take advantage of favourable institutional conditions to achieve their desired outcomes. They ... have to bargain, cajole, and utilize any

knowledge or institutional power beneficial to their cause (Zito, 2000: 11).

Within an enlarged Community, the need for new environmental proposals to obtain concurrent-majority support is liable to trigger more intense interest-led bargaining because of the expanded range of interests present. The task of environmental entrepreneurs will therefore become increasingly demanding as broader concessions are needed for policies to gain acceptance. Whilst leader states will undoubtedly continue to press their national agendas and trade considerations will perpetuate the push-pull dynamic, the complexities of joint decision-making suggest that a more flexible approach to environmental policy will be needed to cope with EU enlargement. Pre-empting such contingencies, the issue of expanded policy-making was one of the few environmental issues debated at the EU's 1996/7 Inter-Governmental Conference. The main theme of these discussions was the need for greater flexibility in standard setting, first, to help manage the lengthy transitions required by some prospective members and, second, as a route forward for environmental policy where not all member states wished to take part (Lowe and Ward, 1998b). Even though the practical and democratic justifications for this tactic are indisputable, it raises the prospect of two or even three-tiers of environmental standards becoming commonplace in EU environmental policy rather than being a short-term phenomenon to accommodate the transition of new member states towards alignment with the *acquis*. This again has implications for maintaining the balance between the EU's trade and environmental policies.

7.3.3 The Deployment of NEPIs in the CEE states

There is little reason to suppose that the CEE accessions will lead to the creation of more NEPIs at EU level. If anything, enlargement will increase the number of economic objections to supranational fiscal measures, whilst more established member states might reasonably question the environmental appropriateness of NEPIs designed with half an eye on the implementation deficiencies of the CEE countries. The EU carbon tax again illustrates this point. Under the Kyoto Protocol the member states agreed a Burden Sharing Agreement equating to an eight per cent reduction in total emissions from a 1990 base year, to be achieved by 2008-12 (table 7.2). Although the CEE states also have targets under Kyoto, their emissions have already declined drastically as a result of economic contraction during their transition to market economies (Tables 7.3). In many cases, major industrial expansion can still be accommodated within these targets and, in fact, the Kyoto negotiations explicitly recognised that the CEE states should be permitted to increase emissions in order to stimulate economic recovery and promote political stability (Grubb *et al.*, 1999). The application of a common carbon/energy tax comparable to that proposed for current member states would therefore be unnecessary to fulfil international obligations and would almost certainly prove economically damaging. Whilst this argument is slightly hypothetical given the failure of the EU carbon/energy tax proposal, it highlights that, for the foreseeable future, the CEE economies will require special consideration away from the mainstream of EU environmental policy (Lowe and Ward, 1998b).

The logic of deploying more national NEPIs in the CEE countries is altogether more complicated, however. On the one hand, it is doubtful whether many businesses in the CEE states could cope with the additional costs imposed by economic instruments. It is often reported that policies to clean up and

Table 7.2   **Distribution of emissions targets under the EU Burden Sharing Agreement**

Country	Internal commitment (% change from 1990 levels)
Austria	-13.0
Belgium	-7.5
Denmark	-21.0
Finland	0
France	0
Germany	-21.0
Greece	+25.0
Ireland	+13.0
Italy	-6.5
Luxembourg	-28.0
Netherlands	-6.0
Portugal	+27.0
Spain	+15.0
Sweden	+4.0
UK	-12.5

Source: Grubb *et al.* (1999:123)

protect the environment are dependant on countries creating sufficient surpluses of resources to devote to environmental protection (Joyce, 2002). Equally, if attempts are made to expand the use of fiscal instruments before adequate resources, legislative frameworks, and enforcement capacity are in place, this could be to the detriment of both environmental and economic policy (EAP Task Force, 1995). Such views tended to prevail in the early stages of CEE transitions, when environmental issues were not perceived as a priority. For example, the Polish Ministry of the Environment encountered almost unanimous opposition when it attempted to introduce a four per cent *ad valorem* fuel charge with the revenue earmarked for environmental protection. Manser (1993: 93) reports Solidarity's official opinion on the proposal thus; 'while the Union is for environmental protection, it will not approve any such burden laid on an impoverished society.' Plans in the early 1990s for economic instruments covering waste, air pollution, the use of underground water, mining, noise, and environmentally benign and undesirable goods in the Czech Republic were also blocked except where links to Western practice could be established (Manser, 1993).

On the other hand, all environmental policies have cost implications and

Table 7.3 Kyoto targets of the EU and CEE states from base year (1990 unless stated)

	Commitment (% from 1990 levels)	1995 emissions (% from base year)
EU	-8	-3.4
Bulgaria (1988)	-8	-35.7
Czech Republic	-8	-21.5
Estonia	-8	-44.4
Hungary (1985-7)	-6	-24.0
Latvia	-8	-46.2
Lithuania	-8	n.a.
Poland (1989)	-6	-22.2
Romania (1989)	-8	-38.1
Slovak Republic	-8	-25.2
Slovenia	-8	n.a.

Source: Grubb *et al.* (1999:118)

the overwhelming evidence is that NEPIs achieve equal or higher standards of environmental protection at lower cost than command-and-control. An accelerated shift towards flexible policy instruments in the CEE countries might therefore mitigate the costs of essential environmental policies and help to reduce implementation problems by integrating environmental factors into market economies as they develop rather than being introduced in a *post hoc* fashion. Furthermore, as Klarer and Francis (1997) point out, the CEE countries can draw upon the experience and technologies of the West and many synergies can be gained from interaction with the EU in smoothing the transition to new forms of environmental regulation. Overall, therefore, NEPIs can further augment the economic benefits to be gained from alignment with the environmental *acquis* (CEC, 1997b).

The ability of CEE economies to develop the market disciplines and implementation capacity needed for the effective operation of NEPIs is nonetheless a major concern. As noted earlier, whilst the Commission currently considers that the Czech Republic, Estonia, Hungary, Poland and Slovenia now have functioning market economies, the ability of many candidate countries to cope with competitive pressures in the internal market is still questionable (CEC, 1999b). The Czech Republic and Slovenia have made the greatest progress against this standard, Hungary and Poland are proceeding apace, but Estonia still has considerable work to complete. Yet the UK's early experiences with the PRN system demonstrate that the mere existence of a mature market economy does not guarantee the success of NEPIs, since free markets do not, of their own volition, take into account the externality costs of environmental degradation. In order for this to happen, governments must also have the capacity to regulate market behaviours that cause environmental neglect.

The extent to which this precondition can currently be met by the CEE governments is debatable, as legal, institutional and administrative weaknesses have created additional obstacles to the effective implementation of environmental policies (Barnes and Barnes, 1999). The environmental impact of industrial activity is still severely under-regulated in most CEE countries and competitive pressures caused by privatisation have militated against the internalisation of environmental costs (Saiko, 1998). Endemic corruption in some national authorities has also militated against effective enforcement (CEC, 1999b)[6]. As Manser (1993: 93) suggests, in many CEE states the introduction of NEPIs has been 'less of an issue than the need for tough enforcement by a competent local ecological "police".'

More recent reports suggest that the CEE governments are making progress in incorporating NEPIs into national environmental policies, however.

Economic instruments have been introduced in various countries covering motor fuels, other energy sources, air emissions, transport, waste, water quality, mining, biodiversity, and nature protection (see Table 7.4) (Klarer *et al.*, 1999a). Nevertheless, the distribution and rates of taxes vary widely between countries

Table 7.4 Overview of selected environmental taxes and charges in the CEE states

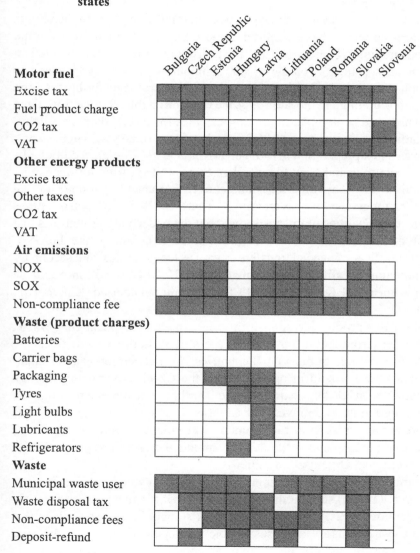

Source: Klarer *et al.* (1999a: pages not numbered)

and it is noteworthy that, by 1999, economic instruments for packaging waste had only been introduced in Estonia, Hungary and Latvia, whilst carbon dioxide charges only existed in Slovenia. Perchards (1998) also report that measures to control packaging waste had been instigated in the Czech Republic, Estonia, Hungary, Latvia, and Slovakia by the late 1990s, some of which incorporate producer responsibility agreements and recovery fees for packaging waste.

Klarer *et al.* (1999b) report that positive experiences with tradable permits in Western countries have inspired some CEE governments to examine emissions trading schemes. In 1991 the Polish government implemented an experimental scheme for volatile organic compounds and although inadequate regulation led to a lack of trading, the experience has prompted further research on emissions trading. In the late 1990s the Czech Republic, Latvia, Lithuania and Poland were contemplating trading schemes for carbon dioxide, sulphur and volatile organic compounds, and a marketable permit system in the water sector was under preparation in Lithuania (Klarer *et al.*, 1999a). Additionally, the CEE countries and New Independent States (NIS) of the former Soviet Union are at the heart of discussions concerning an international trading mechanism for greenhouse gas emissions. There are still many open questions about the technicalities of this instrument but the underlying principle is that the CEE and NIS could sell their excess 'hot air' to countries like the USA, creating significant financial transfers as well as technological and secondary environmental benefits. It is estimated that US$10-20 billion could be generated during the Kyoto Protocol's first commitment period (Klarer *et al.*, 1999b).

Aside from producer responsibility agreements for packaging waste, voluntary environmental agreements are still virtually absent from the regulatory armouries of most CEE countries. One exception is a voluntary agreement between landfill operators and the Czech government, though environmental NGOs have expressed the fear that such agreements might be exploited by industry to prevent the adoption of essential regulation. They therefore argue that support for voluntary instruments should be contingent on their transparency and verification by an independent third party (Slovak Non-Government Organisations, 2000).

Summing up the prospects for environmental policy in the CEE countries, Klarer *et al.* (1999b) suggest that significant benefits can be gained from the deployment of NEPIs as a complement to existing policies. They claim that economic instruments are key to sustainable development in the region because they encourage the integration of environmental concerns into the economic incentive structures of the new market economies, while support for the polluter and user pays principles is essential for determining past and current environmental liabilities. Finally, they highlight the economic benefits of NEPIs for the CEE economies in terms of improving innovation and competitiveness, as well as their potential to raise revenue for environmental investments. Their proposed strategy for environmental policy in the CEE countries therefore consists of an accelerated uptake of economic instruments based upon a systematic evaluation of their current and potential uses, clear identification of policy objectives, and early stakeholder involvement to ensure economic instruments are effectively designed and implemented. They also recommend that consideration be given to the formation of green budget commissions to oversee the strategic integration of environmental concerns into fiscal policy, taking into account Western experiences (Klarer *et al.*, 1999b).

Although these arguments are based primarily on the internal benefits of NEPIs for the CEE countries, EU accession will also be a key factor in their deployment in the region. Notwithstanding subsidiarity arguments, the alignment of legal systems may spill over into pressure for the harmonisation of implementation methodologies as EMU increases states' sensitivity towards disparities in economic instruments. Agenda setting by the Commission and leader states may therefore pressurise new states to accept greater innovation than might emerge through purely nationally driven policies. In the final analysis, however, the institutions, legal frameworks and implementation procedures used to enforce environmental policy in the CEE countries will need to be strengthened substantially to support the wider use of flexible policy instruments. Whilst it is essential that environmental resources are properly valued by the CEE economies, it seems prudent that all parties in the accession process pursue capacity building as a first priority before embarking on ambitious programmes of NEPI deployment.

7.4 Conclusions

There can be little doubt that new environmental policy instruments have helped to push implementation towards the top of the agenda of EU environmental policy and have provided decision-makers with a more extensive armoury of mechanisms with which to achieve their objectives. The intention of this book has been to provide a critical commentary on the translation of NEPIs from theory into practice. From the evidence presented it is apparent that NEPIs are making a genuine contribution towards reducing the implementation deficit in the member states by providing innovative means for incorporating environmental considerations into the decisions of the market.

Although command-and-control regulation would outwardly appear to offer a more straightforward route towards achieving sustainability, experience has shown that such devices do not have the same integrative capacity as NEPIs. NEPIs have also helped to mitigate the costs of environmental policy, addressing the point that the long-term success of environmental policies ultimately depends on their affordability. By developing mechanisms that seek to maximise social welfare through the balancing of environmental, social and economic priorities, economists have made significant progress in persuading policy-makers that environmental protection is no longer an unaffordable luxury. Finally, suasive NEPIs such as voluntary environmental agreements have done much to nurture co-operation between regulators and regulated in the development of standards, methodologies and implementation time frames.

It is equally apparent that policy-makers are still coming to terms with the design and deployment of NEPIs. The study has revealed few weaknesses in the theories underpinning most NEPIs; rather, the majority of instances where policies have under-performed can be traced to regulatory failures. The first major type of failing is defective instrument design, which most commonly arises when insufficiently clear policy objectives are set by decision-makers. This may lead, *inter alia*, to the creation of inappropriate incentives or misinterpretation of the way markets can respond to particular stimuli. The second source of underperformance relates to application weaknesses, where inadequate monitoring or enforcement mechanisms lead to parties targeted by NEPIs failing to abide by the letter or, in the case of PRNs, the spirit of environmental policies.

NEPIs are nonetheless not a panacea for all environmental problems, particularly those created by demand for price-inelastic commodities or where an immediate reduction or cessation of pollution is necessary, because they contain fewer binding mandates than command-and-control on the behaviour of polluters. Furthermore, under-performance of NEPIs cannot be blamed entirely on administrative shortcomings, as national politicians are democratically obliged to take account of a host of competing priorities when designing and implementing environmental policies. As a result, it is often simply not possible to introduce NEPIs in their textbook form. Such tensions exist at all levels of government but are undoubtedly amplified within the multi-tiered environmental governance of the EU. Not only must NEPIs accommodate competing national prerogatives, they must also co-exist with the over-arching European agendas of economic union, trade harmonisation and subsidiarity. It is not surprising, therefore, that the transition of NEPIs from theory to practice in the EU has involved a significant amount of compromise and policy adaptation.

As a final point, it is worth reflecting on NEPIs in the context of the wider changes that have taken place in EU environmental policy over the last thirty years. Environmental thinking has undergone a major transition during this period as the concept of limits to growth has been progressively replaced by the more eclectic vision of sustainable development (O'Riordan and Voisey, 1998). The transformation in attitudes towards the implementation of environmental policies is no less intriguing. Following an extended period when environmental problems were largely ignored or subordinated behind more immediate priorities such as post-war economic reconstruction and the provision of minimum standards of social welfare, the early EAPs were characterised by command-and-control measures designed to regulate specific forms of degradation. However, policies during this era generally lacked a strategic overview and, as often as not, were initiated primarily to defend the internal market. As the limitations of this piecemeal and techno-centric approach were recognised, EU decision-makers sought to develop a more synoptic approach through such devices as environmental impact assessment. Throughout both periods, the accent of policy intervention nonetheless remained oriented towards standards-based legislation.

The emergence of NEPIs represents an important juncture in EU environmental policy providing the clearest indication to date of European decision-makers' determination to integrate environmental concerns into all spheres of policy and planning. Although this has been a longstanding objective of the EU environmental programme, progress towards this ambition was generally disappointing prior to the advent of NEPIs. NEPIs have also been instrumental in increasing the constituency of environmental policy, as governments have sought to consult more broadly on the design and deployment of new implementing mechanisms. Notwithstanding these encouraging trends, it is also conspicuous that none of these approaches to environmental policy have seriously questioned the appropriateness of continued economic expansion in the EU. Though determined efforts have been made to re-define the EU's political agenda to reflect the values of sustainable development, economic growth remains a key priority for most member states. In practical terms, the main focus of activity has been on developing *new methods* for implementing environmental policies rather than on envisioning *new perspectives* towards the organisation of economic activity. However, if new environmental policy instruments do not enable the member states to make significant progress towards sustainable development then the EU may be forced to initiate a more fundamental debate on its future direction and values.

Notes

[1] If the UK abstained from EU-wide permit trading, this could reduce the economies of scale to be gained from such schemes. However, the UK has been at the vanguard of tradable permits in Europe and, therefore, is unlikely to refrain from trading where there is a clear economic or environmental rationale. Although it might be more costly for companies from other member states to trade with UK firms because of currency exchanges, the impact of non-adoption of the Euro on permit schemes is unlikely to be significant.

[2] According to the current timetable, the first wave of accessions in 2004 will include Cyprus, the Czech Republic, Estonia, Hungary, Latvia, Lithuania, Malta, Poland, the Slovak Republic and Slovenia. The other states being considered for EU membership are Bulgaria, Romania, and Turkey.

[3] The PHARE programme was originally set up to support economic and political transition in Poland and Hungary by means of technical assistance, training, grants and capital investments. It has subsequently been expanded to provide similar assistance to other CEEC. In 1999, it had a budget of Euro 6.693 billion.

[4] However, Jordan (2000) also argues that subsidiarity has not been materially applied in environmental policy to challenge the EU's authority to continue regulating in areas where directives were adopted in a rather *ad hoc* fashion during the 1970s and 1980s. In such areas, Jordan suggests, subsidiarity has become a euphemism for 'better law making' as the EU has sought to develop a more nuanced apportionment of strategic and tactical planning responsibilities between the Commission and the member states.

[5] The Luxembourg compromise arose as a result of the actions of Charles de Gaulle, the French President between 1958 and 1968, who perceived the EEC developing as a Europe of independent states with no surrendering of national sovereignty to supranational institutions. His support for French national interests was exemplified by a crisis which emerged in the 1960s over funding for the Common Agricultural Policy. In 1965 de Gaulle took the decision to boycott the EEC institutions in opposition to the introduction of majority voting, which would have blocked France's interests on agricultural issues. For several months the French observed an 'empty-chair' policy, arguing that each government and not the Community should determine how national interests were to be defined. The crisis ended in early 1966 when the French government forced the Luxembourg Compromise, which stipulated that majority voting would not be used when any member government felt its national interests were being seriously jeopardised by EU policies. However, the principle of qualified majority voting has now been accepted in many areas of environmental policy.

[6] Aside from corruption, Manser (1993) discusses three main obstacles to the effective monitoring and enforcement of environmental policy in the CEE states. First, the decentralisation of monitoring activities compared with the main decision-making apparatus in most CEE countries means that local enforcement offices lack the necessary staff, skills and power to enforce legislation. Second, many industrial plants simply lack the necessary control equipment. Third, some CEE governments have allowed industry effectively to side-step fines or charges in order to protect their country's economic performance in times of recession.

Bibliography

ACP (1998) *Advisory Committee on Packaging report on review of the Producer Responsibility Obligations (Packaging Waste) Regulations*, London: DETR.

ACP (2001) *Report of the task force of the Advisory Committee on Packaging*, London: DEFRA.

Agyeman, J. and Evans, R. (1997) 'Government, sustainability and community,' *Local Environment*, 2 (2), 117-8.

Archer, C. and Butler, F. (1996) *The European Union: structure and process (second edition)*, London: Pinter.

Atkinson, S. and Tietenberg, T. (1991) 'Market failure in incentive-based regulation: the case of emissions trading', *Journal of Environmental Economics and Management*, 21 (1), 17-31.

Bahn, O. (1999) 'Combining policy instruments to curb greenhouse gas emissions,' *ERP Environment International Sustainable Development Conference 1999*, 11-18.

Bailey, I.G. (1999a) 'Flexibility, harmonization and the Single Market in European Union environmental policy: evidence from the Packaging Waste Directive,' *Journal of Common Market Studies*, 37 (4), 549-72.

Bailey, I.G. (1999b) 'Competition, sustainability and packaging policy in the UK,' *Journal of Environmental Planning and Management*, 42 (1), 83-102.

Bailey, I.G. (1999c) 'The development of the Packaging Regulations: a review of the reprocessing industry,' *Wastes Management*, February 1999, 38-9.

Bailey, I.G. (1999d) 'The development of the Packaging Regulations II: the response from packaging producers,' *Institute of Wastes Management Scientific and Technical Review*, September, 10-19.

Bailey, I.G. (2000) 'Principles, policies and practice: evaluating the environmental sustainability of Britain's Packaging Regulations,' *Sustainable Development*, 8 (1), 51-64.

Bailey, I.G. (2002) 'European environmental taxes and charges: economic theory and policy practice,' *Applied Geography*, 22 (3), 235-51.

Baker, S. (1997) 'The evolution of European Union environmental policy,' in S. Baker, M. Kousis, D. Richardson and S. Young (Eds) *The politics of sustainable development: theory, policy and practice within the European Union*, London: Routledge, 91-106.

Baranzini, A., Goldemberg, J. and Speck, S. (2000) 'A future for carbon taxes,' *Ecological Economics*, 32 (3), 395-412.

Barde, J-P. (1997) 'Environmental taxation: experience in OECD countries,' in T. O'Riordan (Ed.) *Ecotaxation*, London: Earthscan, 223-45.

Barnes, P.M. and Barnes, I.G. (1999) *Environmental policy in the European Union*, Cheltenham: Edward Elgar.

Barrett, A. and Lawlor, J. (1997) 'Questioning the waste hierarchy: the case of a region with a low population density,' *Journal of Environmental Planning and Management*, 40 (1), 19-36.

Barrett, A., Lawlor, J. and Scott, S. (1997) *The fiscal system and the polluter pays principle*, Aldershot: Ashgate.

Baumol, W.J. and Oates, W.A. (1979) *Economics, environmental policy, and the quality of life*, Englewood Cliffs: Prentice-Hall.

Baumol, W.J. and Oates, W.E. (1988) *The theory of environmental policy (second edition)*, Cambridge: Cambridge University Press.

Beckerman, W. (1974) *In defence of economic growth*, London: Jonathan Cape.

Beder, S. (1996) 'Charging the earth: the promotion of price-based measures for pollution control,' *Ecological Economics*, 16 (1), 51-63.

Beesley, M.E. and Littlechild, S.C. (1983) 'Privatisation: principles, problems and priorities,' *Lloyds Bank Annual Review*, 149, 1-20.

Beynon, D. (1993) 'Environmentally responsible packaging manufacture,' in G.M. Levy (Ed.) *Packaging in the environment*, London: Blackie Academic and Professional, 87-117.

Bhagwati, J. (1993) 'The case for free trade,' *Scientific American*, 269 (5), 18-23.

Bickerstaffe, J. and Barrett, E. (1993) 'Packaging's role in society,' in G.M. Levy (Ed.) *Packaging in the environment*, Blackie Academic and Professional, 34-52.

Bizer, K. and Jülich, R. (1999) 'Voluntary agreements – trick or treat?' *Journal of European Environmental Policy*, 9 (1), 59-66.

Blackhurst, R. (1994) *Trade and sustainable development principles*, London: International Institute for Sustainable Development.

Blacksell, M. (1994) 'Environmental policies and resource management,' in M. Blacksell and A.M. Williams (Eds) *The European challenge*, Oxford: Oxford University Press, 323-42.

Blacksell, M. (1998) 'Redrawing the political map,' in D. Pinder (Ed.) *The new Europe: economy, society and environment*, Chichester: John Wiley and Sons, 23-42.

Blüdhorn, I., Krause, F. and Scharf, T. (eds) (1995) *The green agenda: environmental politics and policy in Germany*, Keele: Keele University Press.

BMU (Bundesministerium für Umwelt, Naturschutz und Reaktorsicherheit) (2000) *Germany's National Climate Protection Programme*, Berlin: BMU.

BMU (2001) 'Cabinet decides on deposit on drinks cans', 30 May 2001, http://www.bmu.de/english/fset800.htm

Bohm, P. (1997) 'Environmental taxation and the double dividend: fact or fallacy?' in T. O'Riordan (Ed.) *Ecotaxation*, London: Earthscan, 106-24.

Böhmer-Christiansen, S. (1994) 'Politics and environmental management,' *Journal of Environmental Planning and Management*, 37 (1), 69-85.

Börkey, P. and Lévêque, F. (2000) 'Voluntary approaches for environmental protection in the European Union – a survey,' *European Environment*, 10 (1), 35-54.

BRC (1998) *Submission to the Packaging Advisory Committee review of the Producer Responsibility Obligations (Packaging Waste) Regulations 1997*, London: BRC.

Brisson, I. (1993) 'Packaging waste and the environment: economics and policy,' *Resources, Conservation and Recycling*, 8 (3-4), 183-292.

British Broadcasting Corporation (2000) 'UK fuel tax: the facts,' http://news.bbc.co.uk/hi/english/in_depth/world/2000/world_fuel_crisis/newsid_933000/933648.

Bulmer, S. (1983) 'Domestic politics and European Community policy making,' *Journal of Common Market Studies*, 21 (4), 349-63.

Bundesverband der Deutschen Industrie (2000) *Agreement on climate protection between the government of the Federal Republic of Germany and German business*, Berlin: *Bundesverband der Deutschen Industrie*.

Capros, P. and Mantzos, L. (2000) *The economic effects of EU-wide industry-level emission trading to reduce greenhouse gases: results from PRIMES energy systems model*, Athens: Institute of Communication and Computer Systems, University of Athens.

Carlsson, F. (2000) 'Environmental taxation and strategic commitment in duopoly models,' *Environmental & Resource Economics*, 15 (3), 243-256.

Carraro, C. (2001) 'Environmental technological innovation and diffusion,' in H. Folmer, H.L. Gabel, S. Gerking and A. Rose (Eds) *Frontiers in environmental economics*, Cheltenham: Edward Elgar, 342-70.

CEC (1984) *Ten years of Community environment policy*, Brussels: CEC.

CEC (1987) *The state of the environment in 1986*, Brussels: CEC.

CEC (1992) *Towards Sustainability: a European Community programme of policy and action in relation to the environment and sustainable development*, COM (92) 93 final: Brussels.

CEC (1993) *General Report on the activities of the European Communities*, General Report 1993, Brussels: CEC.

CEC (1994) *Potential benefits of integration of environmental and economic policies*, Brussels: Graham and Trotman and CEC.

CEC (1996a) *Communication on the review of the Community strategy for waste management*, COM 96 (399) final: Brussels.

CEC (1996b) *Communication from the Commission to the Council and the European Parliament on environmental agreements*, COM (96) 561 final: Brussels.

CEC (1997a) 'The Commission proposes a common system for the taxation of energy products', *European Commission Spokesman's Service,* 12 March 1997: Brussels.

CEC (1997b) *Agenda 2000: communication for a wider and stronger Europe (including the Commission opinions on the applicant states from Central and Eastern Europe)*, COM (97) 2000 final: Brussels.

CEC (1998a) *Fifteenth annual report on monitoring the application of Community Law*, COM (98) 317 final: Brussels.

CEC (1998b) 'Commission decides further action against Greece, Ireland and Luxembourg on packaging waste,' *European Commission press release*, IP/98/579, 30 June 1998.

CEC (1998c) 'Commission decides action against the United Kingdom on packaging waste,' *DGXI Press Release*, ip/98/330.

CEC (1999a) *Sixteenth report on monitoring the application of Community law*, COM (99) 301 final: Brussels.

CEC (1999b) *Composite paper: reports on progress towards accession by each of the candidate countries*, COM (1999) 500 final: Brussels.

CEC (2000a) *Adapting the institutions to make a success of enlargement: Commission opinion*, COM (2000) 34: Brussels.

CEC (2000b) *EU policies and measures to reduce greenhouse gas emissions: towards a European climate change programme (ECCP)*, COM (2000) 88 final: Brussels.

CEC (2000c) *Cost-efficiency of packaging recovery systems: the case of France, Germany, the Netherlands and the United Kingdom*, Brussels: CEC and Taylor Nelson Sofres Consulting.

CEC (2000d) *Guide to the approximation of European Union environmental legislation. part one: introduction to the approximation of environmental legislation*, http://europa.eu.int/comm/environment/guide/part1.htm

CEC (2001a) *Sixth Environmental Action Programme. Environment 2010: our future, our choice*, COM (2001) 31 final: Brussels.

CEC (2001b) *The future of the Common Fisheries Policy*, COM (2000) 135: Brussels.

CEC (2001c) *Making a success of enlargement: strategy paper and report of the European Commission on the progress towards accession by each of the candidate countries*, COM (not numbered) final: Brussels.

Collins, K. and Earnshaw, D. (1993) 'The implementation and enforcement of European Community environmental legislation,' in D. Judge (Ed.) *A green dimension for the European Community: political issues and processes*, London: Frank Cass, 213-49.

Cooper, J. (2000) 'EU Ponders over type of target to take forward,' *Materials Recycling Week*, 28 January, 8-9.

Cowgill, A. (1992) *The Maastricht Treaty in perspective*, Stroud: British Management Data Foundation.

Cummings, R., McKee, M. and Taylor, L. (2001) 'To whisper in the ears of princes: laboratory economic experiments and environmental policy,' in H. Folmer, H.L. Gabel, S. Gerking and A. Rose (Eds) *Frontiers of environmental economics*, Cheltenham: Edward Elgar, 121-47.

Daily Telegraph (1997) 'Brussels flowerpot men fail to amuse M&S,' *30/8/97, Business News*, 1.

Daly, H.E. (1992) *Steady state economics*, London: Earthscan.

Daly, H.E. (1993) 'The perils of free trade,' *Scientific American*, 269 (5), 24-9.

Daly, H.E. and Cobb, J. (1990) *For the common good*, London: Green Point.

Defeuilley, C. and Godard, O. (1997) 'The economics of packaging waste recycling in France: institutional framework and technological adoption,' *International Journal of Environment and Pollution*, 7 (4), 538-46.

DEFRA (2001) *Consultation paper on recovery and recycling targets for packaging waste in 2002: the Producer Responsibility Obligations (Packaging Waste) Regulations*, London: DEFRA.

Debelke, J. and Bergman, H. (1998) 'Environmental taxes and charges in the EU,' in J. Golub (Ed.) *New instruments for environmental policy in the EU*, London: Routledge.

Demmke, C. (1994) *Die Implementation von EG-Umweltpolitik in den Mitgliedstaaten*, Baden-Baden: Nomos.

Demmke, C. (Ed.) (1997) *Managing European environmental policy: the role of the member states in the policy process*, Maastricht: European Institute of Public Administration.

DETR (1997a) *The Producer Responsibility Obligations (Packaging Waste) Regulations 1997: user's guide*, Norwich: The Stationery Office.

DETR (1997b) *A packaging catalogue to support the Producer Responsibility Obligations (Packaging Waste) Regulations 1997: ready reckoner*, Norwich: The Stationery Office.

DETR (1998a) *The Packaging (Essential Requirements) Regulations 1998*, Statutory Instrument 1998 No. 1164, London: HMSO.

DETR (1998b) *Review of the Producer Responsibility Obligations (Packaging Waste) Regulations 1997: a consultation paper*, Norwich: The Stationery Office.

DETR (1998c) *Less waste more value: a consultation paper on the waste strategy for England and Wales*, Norwich: HMSO.

DETR (1999a) *Increasing recovery and recycling of packaging waste in the United Kingdom - the challenge ahead: a forward look for planning purposes*, London: The Stationery Office.

DETR (1999b) *A way with waste: a draft strategy for England and Wales, Parts 1 and 2*, Norwich: HMSO.

DETR (1999c) *Consultation paper on changes to the percentage activity obligations and other matters*, Norwich: HMSO.

DETR (1999d) *The Producer Responsibility Obligations (Packaging Waste) (Amendment) Regulations 1999*, London: The Stationery Office.

DETR (1999e) *A summary of the response to the climate change consultation*, Norwich: HMSO.

DETR (2000a) *Consultation paper on implementation of Council Directive 1999/31/EC on the Landfill of Waste*, Norwich: HMSO.

DETR (2000b) *A greenhouse gas emissions trading scheme for the United Kingdom: consultation document*, Norwich: HMSO.

Dimitrakopoulos, D. and Richardson, J. (2001) 'Implementing EU public policy', in J. Richardson (ed.) *European Union: power and policy-making*, London: Routledge, 335-56.

DoE (1993) *Making markets work for the environment*, London: HMSO.

DoE (1996a) *The Producer Responsibility Obligations (Packaging Waste) Regulations: a consultation paper*, London: HMSO.

DoE (1996b) *Draft Producer Responsibility Obligations (Packaging Waste) Regulations 1996: draft user's guide*, London: HMSO.

DoE (1997) *The Producer Responsibility Obligations (Packaging Waste) Regulations 1997*, Statutory Instrument 1997 No. 648, London: HMSO.

Dovers, S., Norton, T. and Handmer, J. (1996) 'Uncertainty, ecology, sustainability, and policy,' *Biodiversity and Conservation*, 5 (10), 1143-67.

Dovers, S., Norton, T. and Handmer, J. (2001) 'Ignorance, uncertainty and ecology,' in J. Handmer, T. Norton and S. Dovers (Eds) *Ecology, uncertainty and policy: managing ecosystems for sustainability*, Harlow: Prentice Hall, 1-25.

Dovers, S.R. (1999) Institutionalising ecologically sustainable development: promises, problems and prospects,' in K.J. Walker and K. Crowley (Eds) *Australian environmental policy 2: studies in decline and devolution*, Kensington: New South Wales University Press, 205-23.

DSD (1998a) *Packaging recycling: techniques and trends*, Bonn: DSD.

DSD (1998b) *Chronicle of closing the loop: farewell to the throwaway society*, Bonn: DSD.

DSD (1998c) *The Green Dot in Europe: we're in*, Cologne: DSD.

DSD (1999a) *Annual report 1998*, Bonn: DSD.

DSD (1999b) *Mass flow verification 1998*, Bonn: DSD.

DSD (2000) *Packaging recycling: sorting, processing, recycling*, Bonn: DSD.

DSD (2001a) *Annual report 2000*, Bonn: DSD.

DSD (2001b) 'Collection zeal unflagging: The Dual System once again fulfils all targets in the year 2000,' http://www.gruener-punkt.de/en/index.

DTI/DoE (1991) *Economic instruments and recovery of resources from waste*, London: HMSO.

DTI/DoE (1992) *Deposit/refund systems for beverage containers and batteries*, London: HMSO.

EAP Task Force-OECD (1995) *EAP Task Force report to the Sofia Ministerial Conference*, Paris: OECD.

Ebreo, A., Hershey, J. and Vining, J. (1999) 'Reducing solid waste: linking recycling to environmentally responsible consumerism,' *Environment and Behaviour*, 31 (1), 84-106.

EEA (2000) *Environmental taxes: recent developments in tools for integration*, Copenhagen: EEA.

Egan, M. (1997) 'Modes of business governance: European management styles and corporate cultures,' *Western European Politics*, 20 (2), 1-21.

Ehrlich, P. (1971) *The population bomb*, London: Ballatine-Friends of the Earth-Pan.

Eichstädt, T., Carius, A. and Krämer, R.A. (1999) 'Producer responsibility within policy networks: the case of German packaging policy,' *Journal of Environmental Planning and Policy*, 1 (1), 133-53.

Ekins, P. (1993) ' "Limits to growth" and "sustainable development": grappling with ecological realities,' *Ecological Economics*, 8 (3), 269-288.

Ekins, P. (1997) 'On the dividends from environmental taxation,' in T. O'Riordan (Ed.) *Ecotaxation*, London: Earthscan, 125-62.

Ekins, P. (1999) 'European environmental taxes and charges: recent experience, issues and trends,' *Ecological Economics*, 31 (1), 39-62.

Ekins, P. and Speck, S. (2000) 'Proposals of environmental fiscal reforms and the obstacles to their implementation,' *Journal of Environmental Policy and Planning*, 2 (1), 93-114.

Elliot, S.J. (1998) 'A comparative analysis of public concern over solid waste incinerators,' *Waste Management and Research*, 16 (4), 351-64.

ENDS (1995a) 'Crunch time for industry's packaging waste plan,' *ENDS Report 246*, July 1995, 31.

ENDS (1995b) 'Rift over packaging waste still to be bridged,' *ENDS Report 245*, June 1995, 37-8.

ENDS (1995c) 'New approach to obligation for packaging waste recovery,' *ENDS Report 248*, September 1995, 30-1.

ENDS (1995d) 'Last chance for industry's packaging waste plan,' *ENDS Report 247*, August 1995, 35-6.

ENDS (1996) 'Hubbub over Packaging Regulations,' *ENDS Report 261*, October 1996, 22-6.

ENDS (1997a) 'More EU law infringement cases announced,' *ENDS Environment Daily*, 19 December 1997.

ENDS (1997b) 'Sweden may oppose German packaging law,' *ENDS Environment Daily*, 24 February 1997.

ENDS (1997c) 'EU threatens German packaging law,' *ENDS Environment Daily*, 14 March 1997.

ENDS (1998a) 'German government tries again on packaging', *ENDS Environment Daily*. 10 June 1998.

ENDS (1998b) 'German packaging law revision nearly finalised', *ENDS Environment Daily*, 12 June 1998.

ENDS (1998c) 'Packaging recovery: a faltering UK experiment with market mechanisms,' *ENDS Report 277*, February 1998, 17-20.

ENDS (1998d) 'Agency makes a mess of waste,' *ENDS Report 280*, May 1998, 25-8.

ENDS (1998e) 'Agencies struggle to enforce Packaging Regulations,' *ENDS Report 282*, July 1998, 17.

ENDS (1998f) 'Packaging Regulations give Agency compliance headache,' *ENDS Report 279*, April 1998.

ENDS (1999a) 'First prosecution under Packaging Regulations,' *ENDS Report 292*, 55.

ENDS (1999b) 'The second conviction under Packaging Regulations,' *ENDS Report 293*, 51.

ENDS (2000a) 'EU countries continuing to flout green laws,' *ENDS Environment Daily*, 5 July 2000.

ENDS (2000b) 'First ever fine imposed on an EU member state,' *ENDS Environment Daily*, 4 July 2000.

ENDS (2000c) 'German drinks packaging talks break down', *ENDS Environment Daily*, 14 June 2000.

ENDS (2000d) 'Commission tackles EU law infringements', *ENDS Environment Daily* 5, July 2000.

ENDS (2000e) 'Hundreds of packaging free riders,' *ENDS Report 306*, 29.

Environment Agency (1998a) *Public register of obligated companies under the 1997 Packaging Waste Regulations*, Environment Agency database.

Environment Agency (1998b) *Public register of compliance schemes under the 1997 Packaging Waste Regulations*, Environment Agency database.

Environment Agency (1998c) *Public register of accredited reprocessors under the 1997 Packaging Waste Regulations*, Environment Agency database.

EUR-OP News (2000) 'EU Law not properly implemented,' *EUR-OP News: Information from the European Communities Publications Office*, 2 (2000) 5.

European Parliament (1993) *Report of the Committee on the Environment, Public Health and Consumer Protection*, Doc. A3-0174/93.

European Parliament (1994) 'Packaging and packaging waste' *European Parliamentary Debate*, No 3-448/10-17.

Fenton, R. and Hanley, N. (1995) 'Economic instruments and waste minimization: the need for discard-relevant and purchase-relevant instruments,' *Environment and Planning A*, 27 (8), 1317-28.

Fenton, R.W. and Sinclair, A.J. (1996) 'Towards a framework for evaluating packaging stewardship programmes,' *Journal of Environmental Planning and Management*, 39 (4), 507-27.

Flanderka, F. (1998) *Verpackungsverordnung: Kommentar*, Heidelberg: Müller.

Folmer, H., Gabel, H.L. and Opschoor, H. (1995) *Principles of environmental and resource economics*, Cheltenham: Edward Elgar.

Folmer, H., Gabel, H.L., Gerking, S. and Rose, A. (Eds) (2001) *Frontiers of environmental economics*, Cheltenham: Edward Elgar.

Gago, A. and Labandeira, X. (2000) 'Towards a green reform model,' *Journal of Environmental Policy and Planning*, 2 (1), 25-37.

Gardner, N. (1996) *A guide to United Kingdom and European competition policy (second edition)*, Basingstoke: Macmillan.

Gee, D. (1997) 'Economic tax reform in Europe: opportunities and obstacles,' in T. O'Riordan (Ed.) *Ecotaxation*, London: Earthscan, 81-105.

Gee, D. and von Weizsäcker, E.U. (1994) 'Shifting the tax burden from economic goods to environmental bads,' *Public Policy Review*, 2 (1), 27-30.

Goddard, H.C. (1995) 'The benefits and costs of alternative solid waste management policies,' *Resources, Conservation and Recycling*, 13 (3-4), 183-213.

Goldman, B.A. (1996) 'What is the future of environmental justice?' *Antipode*, 28 (2), 122-41.

Golub, J. (1996) 'State power and institutional influence in European integration: lessons from the packaging waste directive,' *Journal of Common Market Studies*, 34 (3), 313-39.

Golub, J. (Ed.) (1998) *New instruments for environmental policy in the EU*, London: Routledge.

Goodman, S.F. (1996) *The European Union*, Basingstoke: Macmillan.

Gouldson, A. and Murphy, J. (1996) 'Ecological modernization and the European Union,' *Geoforum*, 27 (1), 11-21.

Green, P. (1994) 'Subsidiarity and European Union: beyond the ideological impasse?' *Policy and Politics*, 22 (4), 287-300.

Grubb, M., Vrolijk, C. and Brack, D. (1999) *The Kyoto Protocol: a guide and assessment*, London: The Royal Institute of International Affairs.

Haas, P.M. (1989) 'Do regimes matter? Epistemic communities and Mediterranean pollution control,' *International Organization*, 43 (3), 377-403.

Haas, P.M. (1992) 'Introduction: epistemic communities and international policy co-ordination,' *International Organization*, 46 (1), 1-35.

Hagengut, C. (1997) *Packaging recovery and recycling in Europe: implementation of the European Directive on packaging and packaging waste*, Bonn: Intec Consulting.

Hahn, R.W. (1989) 'Economic prescriptions for environmental problems: how the patient followed the doctor's orders,' *Journal of Economic Perspectives*, 3 (2), 95-114.

Hahn, R.W. (1993) 'Getting more environmental protection for less money: a practitioner's guide,' *Oxford Review of Economic Policy*, 9 (4), 112-23.

Hahn, R.W. (2000) 'The impact of economics on environmental policy,' *Journal of Environmental Economics and Management*, 39 (3), 375-99.

Haigh, N. (1998) 'Including the concept of sustainable development into the Treaties of the European Union,' in T. O'Riordan and H. Voisey (Eds) *The transition to sustainability: the politics of Agenda 21 in Europe*, London: Earthscan, 64-75.

Haigh, N. and Lanigan, C. (1995) 'The impact of the EU on UK environmental policy making,' in T. Gray (Ed.) *UK environmental policy in the 1990s*, Basingstoke: MacMillan, 18-37.

Halkos, G.E. (1996) 'Implementing optimal sulphur abatement strategies in Europe,' *Water, Air, and Soil Pollution*, 87, 329-44.

Handelsblatt (1997) '*Der Grüne Punkt ist aus den roten Zahlen,' Handelsblatt*, 25 July 1997.

Hanley, N. and Slark, R. (1994) 'Cost-benefit analysis of paper recycling: a case study and some general principles,' *Journal of Environmental Planning and Management*, 37 (2), 189-97.

Haverland, M. (1999) *National autonomy, European integration and the politics of packaging waste*, Amsterdam: Thela Thesis.

Haverland, M. (2000a) 'National adaptation to European integration: the importance of institutional veto points,' *Journal of Public Policy*, 20 (1): 83-103.

Haverland, M. (2000b) 'Impact of the European Union on national policies: the state of the art', *Netherlands Institute of Government Working Papers* 2000-4.

Hayes-Renshaw, F. and Wallace, H. (1995) 'Executive power in the European Union: the functions and limits of the Council of Ministers,' *Journal of European Public Policy*, 2 (4), 559-82.

Helm, D. (1993) 'The assessment: reforming environmental regulation in the UK,' *Oxford Review of Economic Policy*, 9 (4), 1-13.

Helm, D. (1998) 'The assessment: environmental policy - objectives, instruments and institutions,' *Oxford Review of Economic Policy*, 14 (4), 1-19.

Heyes, A.G. (1998) 'Making things stick: enforcement and compliance,' *Oxford Review of Economic Policy*, 14 (4), 50-63.

Heyvaert, V. (2001) 'Balancing trade and environment in the European Union: proportionality substituted,' *Journal of Environmental Law*, 13 (3), 392-407.

Hildebrand, P. (1993) 'The European Community's environmental policy 1957-1992,' *Journal of Environmental Politics*, 1 (4), 13-44.

Hill, K.E. (1997) 'Supply-chain dynamics, environmental issues and manufacturing firms,' *Environment and Planning A*, 29 (7), 1257-74.

HM Customs and Excise (2000) 'HM Customs and Excise REV/C&E 4/00: Climate Change Levy,' http://www.hmce.gov.uk/notices/pnj4-00.htm.

HM Treasury (1999) 'Further details announced on the Climate Change Levy,' *HM Treasury Press Office*, http://www.hm-treasury.gov.uk.

Hoel, M. (2001) 'International trade and the environment: how to handle carbon leakage,' in H. Folmer, H.L. Gabel, S. Gerking and A. Rose (Eds) *Frontiers of environmental economics*, Cheltenham: Edward Elgar, 176-91.

Höreth, M. (1999) 'No way out for the Beast? The unsolved legitimacy problem of European governance,' *Journal of European Public Policy*, 6 (2), 249-68.

House of Commons (1997) 'Packaging waste,' *House of Commons Hansard Debates for 3 March 1997 (pt 9)*, London: The Stationery Office.

House of Lords (1992) *Ninth report of the House of Lords Select Committee on the European Communities, interpretation and enforcement of environmental legislation*, HL Paper 53.1, 10 March, paragraph 67, London: HMSO.

House of Lords (1996) *Select Committee on the European Communities, Eleventh Report, Session 1995-6*, HL Paper 29, London: HMSO.

Howe, C.W. (1996) 'Making environmental policy in a federation of states,' in J.B. Braden, H. Folmer and T.S. Ulen (Eds) *Environmental policy with political integration: The European Union and United States*, Cheltenham: Edward Elgar, 21-34.

Huppes, G., van der Voel, E., van der Naald, W.G.H., Vonkerman, G.H. and Maxson, P. (1992) *New market-oriented instruments for environmental policy*, London: Graham and Trotman for the CEC.

Jacobs, M. (1991) *The green economy: environment, sustainable development and the politics of the future*, London: Pluto Press.

Jacquemin, A. and Wright, D. (1994) 'Corporate strategies and European challenges post-1992,' in S. Bulmer and A. Scott (Eds) *Economic and political integration in Europe: internal dynamics and global context*, Oxford: Blackwell, 219-31.

Jasanoff, S. (1996) 'Beyond epistemology: relativism and the engagement in the politics of science,' *Social Studies of Science*, 26 (2), 393-417.

Jones, C.O. (1997) *An introduction to the study of public policy (third edition)*, Belmont: Wadsworth Publishing.

Jones, E. (1999) 'Competitive and sustainable growth: logic and inconsistency,' *Journal of European Public Policy*, 6 (3), 359-75.

Jordan, A. (1999) 'The implementation of EU environmental policy: a policy problem without a political solution?' *Environment and Planning C*, 17 (1), 69-90.

Jordan, A. (2000) 'The politics of multilevel environmental governance: subsidiarity and environmental policy in the European Union,' *Environment and Planning A*, 32 (7), 1307-24.

Jordan, A. (2002) 'Introduction: European Union environmental policy – actors, institutions and policy processes,' in A. Jordan (Ed.) *Environmental policy in the European Union*, London: Earthscan, 1-10.

Jordan, A., Wurzel, R.K.W., Zito, A.R. and Brückner, L. (forthcoming) 'Policy innovation or 'muddling through'? 'New' environmental policy instruments in the United Kingdom,' *Journal of Environmental Politics*, 11 (4).

Jordan, A. and Jeppesen, T. (2000) 'EU environmental policy: adapting to the principle of subsidiarity,' *European Environment*, 10 (2), 64-74.

Joyce, J.K. (2002) *The political economy of the environment*, Cheltenham: Edward Elgar.

Karamanos, P. (2001) 'Voluntary environmental agreements: evolution and definition of a new environmental policy approach,' *Journal of Environmental Planning and Management*, 44 (1), 67-84.

Kaufmann, R. (2001) 'The environment and economic well-being,' in H. Folmer, H.L. Gabel, S. Gerking and A. Rose (Eds) *Frontiers in environmental economics*, Cheltenham: Edward Elgar, 36-56.

Keay-Bright, S. (2000) *A critical analysis of the voluntary fuel agreements, established between the automobile industry and the European Commission, with regard for their capacity to protect the environment*, document no. 2000/021, Brussels: European Environmental Bureau.

Keohane, R.O. and Nye, J.S. (1989) *Power and interdependence*, London: Harper Collins.

Kimber, C. (2000) 'Implementing European environmental policy and the directive on environmental information,' in C. Knill and A. Lenschow (Eds) *Implementing EU environmental policy: new directions and old problems*, Manchester: Manchester University Press, 168-96.

Klarer, J. and Francis, P. (1997) 'Regional overview,' in J. Klarer and B. Moldan (Eds) *The environmental challenge for central European economies in transition*, Chichester: John Wiley and Sons, 1-66.

Klarer, J., McNicholas, J. and Knaus, E-M. (Eds) (1999a) *Sourcebook on economic instruments for environmental policy in Central and Eastern Europe: a regional analysis*, Szentendre: Regional Environmental Centre for Central and Eastern Europe.

Klarer, J., Francis, P. and McNicholas, J. (1999b) *Improving environment and economy: the potential of economic instruments for environmental improvements and sustainable development in countries with economies in transition*, Szentendre: Regional Environmental Centre for Central and Eastern Europe.

Knill, C. and Lenschow, A. (1998) 'Coping with Europeanisation: the impact of British and German administrations on the implementation of EU environmental policy,' *Journal of European Public Policy*, 5 (4), 595-614.

Kohlhaas, M. (2000) 'Ecological tax reform in Germany: from theory to policy,' *Economic Studies Program Series Vol. 6.*, Baltimore: American Institute for Contemporary German Studies, Johns Hopkins University.

Krämer, L. (1991) 'The implementation of Community environmental directives within member states: some implications of the direct effect doctrine,' *Journal of Environmental Law*, 39 (1), 39-56.

Krämer, L. (1996) '*Defizite im Vollzug des EG-Umweltrechts und ihre Ursachen*,' in G. Lübbe-Wolf (Ed.) *Der Vollzug des europäischen Umweltrechts*, Berlin: Erich Schmidt.

Labatt, S. (1991) 'A framework for assessing discretionary corporate performance towards the environment,' *Environmental Management*, 15 (2), 163-78.

Labatt, S. (1997a) 'Corporate responses to environmental issues: packaging,' *Growth and Change*, 28 (1), 67-92.

Labatt, S. (1997b) 'External influences on environmental decision making: a case study of packaging waste reduction,' *The Professional Geographer*, 49 (1), 105-16.

Levenson, H. (1993) 'Municipal solid waste reduction and recycling: implications for federal policymakers,' *Resources, Conservation and Recycling*, 8 (1-2), 21-37.

Lévêque, F. (1995) 'Standards and standards-setting processes in the field of the environment,' in R. Hawkins, R. Mansell and J. Skea (Eds) *Standards, innovation and competitiveness: the politics and economics of standards in the natural and technical environments*, Aldershot: Edward Elgar, 105-21.

Lévêque, F. (1996a) 'The regulatory game,' in F. Lévêque (Ed.) *Environmental policy in Europe: industry, competition and the policy process*, Cheltenham: Edward Elgar, 31-51.

Lévêque, F. (1996b) 'The European fabric of environmental regulations,' in F. Lévêque (Ed.) *Environmental policy in Europe: industry, competition and the policy process*, Cheltenham: Edward Elgar, 9-30.

Liefferink, D. and Andersen, M. (1998) 'Strategies of the "green" member states in EU environmental policy-making,' *Journal of European Public Policy*, 5 (2), 254-70.

Lister, C. (1996) *European Union environmental law: a guide for industry*, Chichester: John Wiley.

Lober, D. (1997) 'Explaining the formation of business-environmentalist collaborations: collaborative windows and the Paper Task Force,' *Policy Sciences*, 30 (1), 1-24.

London, M. & Llamas, C. (1994) 'Packaging laws in France and Germany,' *Journal of Environmental Law*, 6 (1), 1-20.

Lowe, P. and Ward, S. (1998a) 'Britain in Europe: themes and issues in national environmental policy,' in P. Lowe and S. Ward (Eds) *British environmental policy and Europe: politics and policy in transition*, London: Routledge, 3-30.

Lowe, P. and Ward, S. (1998b) 'Conclusions: lessons and prospects,' in P. Lowe and S. Ward (Eds) *British environmental policy and Europe: politics and policy in transition*, London: Routledge, 285-99.

Maddison, D., Pearce, D.W., Johansson, O., Calthrop, E., Litman, T. and Verhoef, E. (1996) *Blueprint 5: the true costs of road transport*, London: Earthscan.

Manser, R. (1993) *The squandered dividend: the free market and the environment in Eastern Europe*, London: Earthscan.

Markyanda, A. (2001) 'Poverty, environment and development,' in H. Folmer, H.L. Gabel, S. Gerking, and A. Rose (Eds) *Frontiers of environmental economics*, Cheltenham: Edward Elgar, 192-213.

Mayhew, A. (1998) *Recreating Europe: the EU's policy towards Central and Eastern Europe*, Cambridge: Cambridge University Press.

Mazey, S. and Richardson, J.J. (1993) *Lobbying in the European Community*, Oxford: Oxford University Press.

McAleavey, P. and Mitchell, J. (1994) 'Industrial relations and lobbying in the structural funds reform process,' *Journal of Common Market Studies*, 31 (2), 237-248.

McLaughlin, A.M. and Greenwood, J. (1995) 'The management of interest representation in the European Union,' *Journal of Common Market Studies*, 33 (1), 143-56.

McLaughlin, A.M., Jordan, G. and Maloney, W.A. (1993) 'Corporate lobbying in the European Community,' *Journal of Common Market Studies*, 31 (2), 191-212.

McQuaid, R.W. and Murdoch, A.R. (1996) 'Recycling policy in areas of low income and multi-storey housing,' *Journal of Environmental Planning and Management*, 39 (4), 545-62.

Meadows, D.H., Meadows, D.L., Randers, J. and Behrens III, W.W. (1972) *The limits to growth*, London: Pan.

Michaelis, P. (1995) 'Product stewardship, waste minimisation and economic efficiency: lessons from Germany,' *Journal of Environmental Planning and Management*, 38 (2), 231-43.

Moravcsik, A. (1991) 'Negotiating the Single European Act: national interests and conventional statecraft in the European Community,' *International Organization*, 45 (1), 651-688.

More, T.A., Averill, J.R. and Stevens, T.H. (1996) 'Values and economics in environmental management: a perspective and critique,' *Journal of Environmental Management*, 48 (2), 397-409.

Morphet, J. and Hams, T. (1994) 'Responding to Rio: a local authority approach,' *Journal of Environmental Planning and Management*, 37 (4), 479-95.

MRW (1998a) 'Changes due,' *Materials Recycling Week*, 18 December, 3.

MRW (1998b) 'Doting on the Green Dot,' *Materials Recycling Week*, 6 November, 22-24.

MRW (1999a) 'New schemes for packaging,' *Materials Recycling Week*, 26 November, 4.

MRW (1999b) 'Plastics industry awaits nourishment of PRNs,' *Materials Recycling Week*, 12 November, 10-3.

MRW (1999c) 'Plastics fear PRN-led dive,' *Materials Recycling Week*, 29 January, 3.

MRW (2000) 'PRN prices set to rise,' *Materials Recycling Week*, 18 August, 3.

Nannerup, N. (1998) 'Strategic environmental policy under incomplete information,' *Environmental & Resource Economics*, 11 (1), 61-78.

Nunan, F. (1999) 'Barriers to the use of voluntary agreements: a case study of the development of Packaging Waste Regulations in the UK,' *European Environment*, 9 (6), 238-48.

O'Brien, M. and Penna, S. (1997) 'European policy and the politics of environmental governance,' *Policy and Politics*, 25 (2), 185-200.

O'Doherty, R. and Bailey, I. (2000) 'Paper chases and glass houses: tradable permits and EU recycling targets,' *IWM Scientific and Technical Review*, 2 (2), 13-17.

O'Riordan, T. and Voisey, H. (1998) 'The political economy of the sustainability transition,' in T. O'Riordan and H. Voisey (Eds*) The transition to sustainability: the politics of Agenda 21 in Europe*, London: Earthscan, 3-30.

OECD (1994*) Environment and taxation: the cases of the Netherlands, Sweden and the United States*, Paris: OECD.

OECD (1996) *Implementation strategies for environmental taxes*, Paris: OECD.

OECD (2002) *Environmentally related taxes database*, http://www.oecd.org/EN/document/0, ,EN-document-8-nodirectorate-no-1-3016-8,00.html.

OJEC (1994a) *European Parliament and Council Directive 94/62/EC of 20 December 1994 on packaging and packaging waste*, OJ No. L 365/10-23.

OJEC (1994b) *Community guidelines on state aid for environmental protection*, OJC 072 , 10.3.94 3-9.

OJEC (2001) *Treaty of Nice. Amending the Treaty on European Union, the Treaties establishing the European Communities and certain related acts*, 10.3.01, EN C80.

Öko Institut (1998) *New instruments for sustainability – the new contribution of voluntary agreements to environmental policy*, Darmstadt: Öko Institut.

Öko-Institut (1999) *Waste prevention and minimisation: final report*, Darmstadt: Öko-Institut report for the Environment Directorate of the European Commission.

Opschoor, J.B. and Vos, H. (1988) *The application of economic instruments for environmental protection in OECD Member States*, Paris: OECD.

Otto, B. (1999) *Recycling and recovery from packaging in domestic waste – LCA type analysis of different strategies, report commissioned by the German and European plastics industry*, Berlin-Brandenburg: DSD and Verband der Chemischen Industrie.

Payne, D.C. (2000) 'Policy-making in nested institutions: explaining the conservation failure of the EU's Common Fisheries Policy,' *Journal of Common Market Studies*, 38 (2), 303-24.

Pearce, D.W. and Barbier, E. (2000) *Blueprint for a sustainable economy*, London: Earthscan.

Pearce, D.W. and Turner, R.K. (1990) *Economics of the natural resources and the environment*, Baltimore: Johns Hopkins University Press.

Pearce, D.W. and Turner, R.K. (1992) 'Packaging waste and the polluter pays principle: a taxation solution,' *Journal of Environmental Planning and Management*, 35 (1), 5-15.

Pearce, D.W. and Turner, R.K. (1993) 'Market-based approaches to solid waste management,' *Resources, Conservation and Recycling*, 8 (1), 63-90.

Pearce, D.W., Barratt, S., Markyanda, A., Barbier, E., Turner, R.K. and Swanson, T. (1995) 'Global warming: the economics of tradeable permits,' in J. Kirkby, P. O'Keefe and L. Timberlake (1995) (Eds) *The Earthscan reader in sustainable development*, London: Earthscan, 348-353.

Pearce, D.W., Markyanda, A. and Barbier, E. (1989) *Blueprint for a green economy*, London: Earthscan.

Pearce, D.W., Turner, R.K., O'Riordan, T., Adger, N., Brisson, I., Brown, K., Dubourg, R., Frankhauser, S., Jordan, A., Maddison, D., Moran, D. and Powell, J. (1993) *Blueprint 3: measuring sustainable development*, London: Earthscan.

Pelletier, L.G., Legault, L.R. and Tuson, K.M. (1996) 'The environmental satisfaction scale: a measure of satisfaction with local environmental conditions and government environmental policies,' *Environment and Behaviour*, 28 (1), 5-26.

Perchards (1998) *Packaging legislation in Europe*, St Albans: Perchards.

Pfander, J.E. (1996) 'Environmental federalism in Europe and the United States: a comparative assessment of regulation through the agency of the member states,' in J.B. Braden, H. Folmer and T.S. Ulen (Eds) *Environmental policy with political integration: the European Union and United States*, Cheltenham: Edward Elgar, 59-131.

Pigou, A.C. (1920) *The economics of welfare*, London; Macmillan

Pinder, J. (1996) 'Economic and monetary union: pillar of a federal polity,' *Publius – the Journal of Federalism*, 26 (4), 123-40.

Porritt, J. (1994) 'Foreword,' in J. Agyeman and B. Evans (Eds) *Local environmental policies and strategies*, London: Longman, xi-xii.

Porter, M. (1990) *The competitive advantage of nations*, London: Macmillan.

Porter, M. (1998) 'Waste management,' in P. Lowe and S. Ward (Eds) *British environmental policy and Europe: politics and policy in transition*, London: Routledge, 195-213.

Porter, M. and van der Linde, C. (1995) 'Toward a new conception of the environment-competitiveness relationship,' *Journal of Economic Perspectives*, 9 (4), 97-118.

Powell, J. and Craighill, A. (1997) 'The UK landfill tax,' in T. O'Riordan (Ed.) *Ecotaxation*, London: Earthscan, 304-20.

Powell, J.C., Craighill, A.L., Parfitt, J.P. and Turner, R.K. (1996) 'A lifecycle assessment and economic evaluation of recycling,' *Journal of Environmental Planning and Management*, 39 (1), 97-112.

Pressman, J. and Wildavsky, A. (1984) *Implementation (second edition)*, Berkeley, CA: University of California Press.

Raymond Communications (1998) *Getting green dotted: the German recycling law explained in plain English*, Riverdale MD: Raymond Communications.

Raymond Communications (2001) 'Companies advised to prepare for Essential Requirements,' *Recycling Policy News Briefs*, 12th December 2001.

Repetto, R., Dower, R.C., Jenkins, R. and Geoghegan, J. (1992) *Green fees: how a tax shift can work for the environment and the economy*, Washington: World Resources Movement.

Saiko, T. (1998) 'Environmental challenges in the new democracies,' in D. Pinder (Ed.) *The new Europe: economy, society, and environment*, Chichester: John Wiley and Sons, 381-99.

Saiko, T. (2000) *Environmental crises*, Harlow: Prentice Hall.

Sbragia, A. (1996) 'Environmental policy: the push-pull of policy-making,' in H. Wallace and W. Wallace (Eds) *Policy-making in the European Union*, Oxford: Oxford University Press, 235-55.

Schelling, T.C. (1983) 'Prices as regulatory instruments,' in T.C. Schelling (Ed.) *Incentives for environmental protection*, London: MIT Press, 1-40.

Scott, A., Peterson, J. and Millar, D. (1994) 'Subsidiarity: a "Europe of the regions" v. the British constitution,' *Journal of Common Market Studies*, 31 (1), 47-67.

Segerson, K. (1996) 'Issues in the choice of environmental policy instruments,' in J. Braden, H. Folmer and T.S. Ulen (1996) *Environmental policy with political integration: the European Union and the United States*, Cheltenham: Edward Elgar, 149-74.

Segerson, K. and Micelli, T.J. (1998) 'Voluntary environmental agreements: good or bad news for environmental protection?' *Journal of Environmental Economics and Management*, 36 (2), 109-30.

Simonsson, E. (1995) *The German Packaging Ordinance and EU environmental policy: subsidiarity, trade barriers and harmonization*, Georgetown: University Center for German and European Studies.

Sinclair, A.J. and Fenton, R.W. (1997) 'Stewardship for packaging and packaging waste: key policy elements for sustainability,' *Canadian Public Administration*, 40 (1), 123-48.

Skea, J. (1995) 'Changing procedures for environmental standards-setting in the European Community,' in R. Hawkins, R. Mansell and J. Skea (Eds) *Standards, innovation and competitiveness: the politics and economics of standards in the natural and technical environments*, Aldershot: Edward Elgar, 122-35.

Slater, M. (1982) 'Political elites, popular indifference and community building,' *Journal of Common Market Studies*, 21 (1,2), 69-87.

Slovak Non-Government Organisations (2000) *Position paper of Slovak NGOs to EU enlargement process in the field of environment*, Bratislava: Slovak Non-Government Organisations.

Smith, S. (1997) 'Environmental tax design,' in T. O'Riordan (Ed.) *Ecotaxation*, London: Earthscan, 21-36.

Spackman, M. (1997) 'Hypothecation: A view from the Treasury,' in T. O'Riordan (Ed.) *Ecotaxation*, London: Earthscan, 45-51.

Staudt, E. (1997) *DSD - an expensive experiment*, Bochum: IAI (*Instituts für angewandte Innovationsforschung*).

Stavins, R.N. and Whitehead, B.W. (1992) 'Dealing with pollution: market-based incentives for environmental protection,' *Environment*, 34 (7), 29-42.

Terence O'Rourke plc/ETSU (2001) *Renewable energy assessment and targets for the South West*, Bristol: Government Office South West.

The Guardian (2000) '91% in favour of protests,' *September 15, 2000*.

Tietenberg, T.H. (1990) 'Economic instruments for environmental regulation,' *Oxford Review of Economic Policy*, 6 (1), 17-33.

Toth, A.G. (1994) 'Is subsidiarity justiciable?' *European Law Review, 19* (3), 268-85.

Tsoukalis, L. (1997) *The New European economy revisited*, Oxford: Oxford University Press.

Turner, R.K. (1993) 'Sustainability: principles and practice,' in R.K. Turner (Ed.) *Sustainable environmental economics and management: principles and practice*, London: Belhaven, 3-36.

Turner, R.K. and Pearce, D.W. (1993) 'Sustainable economic development: economic and ethical principles,' in E.B. Barbier (Ed.) *Economics and ecology: new frontiers and sustainable development*, London: Chapman and Hall, 177-94.

Turner, R.K., Salmons, R., Powell, J. and Craighill, A. (1998) 'Green taxes, waste management and political economy,' *Journal of Environmental Management*, 53 (1), 121-136.

UK Research Office (2002) 'Emissions trading kicks off in the UK,' *UKRO European News*, 5 April 2002, 5-6.

Umweltbundessamt (1991) *Ordinance on the Avoidance of Packaging Waste of 12 June 1991*, Bonn: *Umweltbundessamt*.

Umweltbundessamt (1998) *Verordnung über die Vermeidung und Verwertung von Verpackungsabfällen vom 27 August 1998*, Bonn: *Umweltbundessamt*.

VALPAK (1998) 'Recovery across Europe: activity in the member states,' *Vantage*, Spring 1998, 12-13.

van den Bergh, J.C. (1996) *Ecological economics and sustainable development: theory, methods and applications*, Cheltenham: Edward Elgar.

van der Straaten, J. (1993) 'A sound European environmental policy: challenges, possibilities and barriers,' in D. Judge (Ed.) *A green dimension for the European Community: political issues and processes*, London: Frank Cass, 65-83.

van Kersbergen, K. and Verbeeck, B. (1994) 'The politics of subsidiarity in the European Union,' *Journal of Common Market Studies*, 32 (2), 215-236.

Vogel, D. (1997) 'Trading up and governing across: transnational governance and environmental protection,' *Journal of European Public Policy*, 4 (4), 556-71.

von Wilmowsky, P. (1993) 'Waste disposal in the internal market: the state of play after the ECJ's ruling in the Walloon import ban,' *Common Market Law Review*, 30 (4), 541-570.

Waite, R. (1995) *Household waste recycling*, London: Earthscan.

Walker, K.J. (2001) 'Uncertainty, epistemic communities and public policy,' in J.W. Handmer, T.W. Morton and S.R. Dovers (Eds) *Ecology, uncertainty and policy: managing ecosystems for sustainability*, Harlow: Prentice Hall, 262-90.

Wallace, H. (1996a) 'The institutions of the EU: experience and experiments,' in H. Wallace and W. Wallace (Eds) *Policy-making in the European Union*, Oxford: Oxford University Press, 37-68.

Wallace, H. (1996b) 'Politics and policy in the EU: the challenge of governance,' in H. Wallace and W. Wallace (Eds) *Policy-making in the European Union*, Oxford: Oxford University Press, 3-36.

Wallace, W. (1996) 'Government without statehood: the unstable equilibrium,' in H. Wallace and W. Wallace (Eds) *Policy-making in the European Union*, Oxford: Oxford University Press, 439-60.

Watkins, M. (2000) 'Quarrying and sustainability: a parody of terms?' *ERP Environment International Sustainable Development Research Conference*, University of Leeds, April 2000, 402-9.

Weale, A. (1996) 'Environmental rules and rule-making in the European Union,' *Journal of European Public Policy*, 3 (4), 594-611.

Weale, A. (1999) 'European environmental policy by stealth: the dysfunctionality of functionalism?' *Environment and Planning C*, 17 (1), 37-51.

Weale, A. and Williams, A. (1992) 'Between economy and ecology? The single market and the integration of environmental policy,' in D. Judge (Ed.) *A green dimension for the European Community: political issues and processes*, London: Frank Cass, 45-64.

Welford, R. and Prescott, K. (1994) *European business: an issue-based approach*, Pitman: London.

Whiston, T. and Glachant, M. (1996) 'Voluntary agreements between industry and government - the case of recycling regulations,' in F. Lévêque (Ed.) *Environmental policy in Europe: industry, competition and the policy process*, Cheltenham: Edward Elgar, 143-74.

Wiers, J. (2002) *Trade and environment in the EC and WTO*, Groningen: Europa Law Publishing.

Wilson, D. (1996) 'Stick or carrot? The use of policy measures to move waste management up the hierarchy,' *Waste Management and Research*, 14 (4), 385-398.

Wise, M. and Gibb, R. (1993) *Single market to Social Europe: the European Community in the 1990s*, Harlow: Longman Scientific and Technical.

Wood, D.M. and Yesilada, B.A. (1996) *The emerging European Union*, Harlow: Longman.

Zito, A.R., (2000) *Creating environmental policy in the European Union*, Basingstoke: Macmillan.

Index